ENGINEERING MATHEMATICS AND STATISTICS

ENGINEERING MATHEMATICS AND STATISTICS

Pocket Handbook

Nicholas P. Cheremisinoff, Ph.D.
Paul N. Cheremisinoff, P.E.

TECHNOMIC
PUBLISHING CO., INC.

LANCASTER · BASEL

Engineering Mathematics and Statistics
a TECHNOMIC publication

Published in the Western Hemisphere by
Technomic Publishing Company, Inc.
851 New Holland Avenue
Box 3535
Lancaster, Pennsylvania 17604 U.S.A.

Distributed in the Rest of the World by
Technomic Publishing AG

Printed in the United States of America
10 9 8 7 6 5 4 3 2 1

Main entry under title:
   Engineering Mathematics and Statistics—Pocket Handbook

A Technomic Publishing Company book
Bibliography: p.

Library of Congress Card No. 89-50812
ISBN No. 0-87762-621-9

Contents

Preface

This pocket handbook is intended as a handy reference guide for engineers, scientists and students on widely used mathematical relationships, statistical formulas and problem-solving methods. It is a compilation of useful formulas and generalized problem-solving techniques employed by practitioners in the analysis and interpretation of scientific data and problem solving. Written in short note form, it is intended to provide the user with quick, easy reference to information with ample references provided for further readings. Illustrated examples are included for more involved problem-solving methods. Many of the techniques, particularly those involving data regression and statistical analysis are well suited to adaptation on personal computers. For these methods, more detailed instructions are included to guide and illustrate computer aided problem solving.

Part A
Mathematics

ANGLE BETWEEN, NORM, AND DOT PRODUCT OF VECTORS

Let $\vec{a} = (a_1, a_2, \ldots, a_n)$ and $\vec{b} = (b_1, b_2, \ldots, b_n)$ be two vectors. The norm of $\vec{a}$ is denoted by $|\vec{a}|$ and is calculated by the following formula:

$$|\vec{a}| = \sqrt{u_1^2 + u_2^2 + \ldots + u_n^2}$$

Similarly,

$$|\vec{b}| = \sqrt{b_1^2 + b_2^2 + \ldots + b_n^2}$$

The dot product of $\vec{a}$ and $\vec{b}$ is denoted by $\vec{a} \cdot \vec{b}$ and is calculated by the following formula:

$$\vec{a} \cdot \vec{b} = a_1 b_1 + a_2 b_2 + \ldots + a_n b_n$$

The angle between a and b is denoted by θ and is calculated by the following formula:

$$\theta = \cos^{-1}\left(\frac{\vec{a} \cdot \vec{b}}{|\vec{a}| \cdot |\vec{b}|}\right)$$

AREA OF A TRIANGLE (a, b, c) (refer to Figure A1)

Given three sides of a triangle this program computes the area by the following formula:

$$\text{Area} = \sqrt{s(s - a)(s - b)(s - c)}$$

where $s = \frac{1}{2}(a + b + c)$.

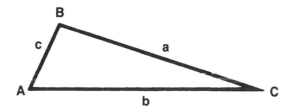

Figure A1. Determination of angle of a triangle (a, b, c).

Example:

Find the area of a triangle with the following three sides:

$$a = 5.31$$
$$b = 7.09$$
$$c = 8.86$$

Solution:

$$Area = 18.82$$

AREA OF A TRIANGLE (a, b, C)
(refer to Figure A2)

Given two sides and an included angle of a triangle this formula computes the area:

$$Area = \tfrac{1}{2} \, ab \sin C$$

Example:

Find the area of the triangle with the following two sides and included angle:

$$a = 5.3174$$
$$b = 7.0898$$
$$C = 45°$$

Solution:

$$Area = 13.33$$

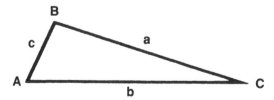

Figure A2. Determination of angle of a triangle (a, b, C).

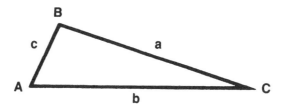

Figure A3. Determination of angle of a triangle (*a, B, C*).

AREA OF A TRIANGLE (*a, B, C*)
(refer to Figure A3)

Given two angles and an included side of a triangle the area is computed by the following formula:

$$\text{Area} = \frac{a^2 \sin B \sin C}{2 \sin (B + C)}$$

Example:

Given the following two angles and included side find the area of the triangle.

$$a = 14.625$$
$$B = 70.54°$$
$$C = 62.96°$$

Solution:

$$\text{Area} = 123.82$$

AREA OF A TRIANGLE [(*x₁, y₁*), (*x₂, y₂*), (*x₃, y₃*)]

Given the coordinates of the vertices of a triangle, the area is found by the following formulas:

Area = ½Determinant of *D* where

$$D = \begin{vmatrix} x_1 & y_1 & 1 \\ x_2 & y_2 & 1 \\ x_3 & y_3 & 1 \end{vmatrix}$$

Therefore,

$$\text{Area} = \frac{1}{2}[x_1(y_2 - y_3) + x_2(y_3 - y_1) + x_3(y_1 - y_2)]$$

Example:

Find the area of the triangle with the following *x-y* coordinate vertices:

$$(0,0)$$
$$(4,0)$$
$$(4,3)$$

Solution:

$$\text{Area} = 6$$

CIRCLE DETERMINED BY THREE POINTS

Let (x_1, y_1), (x_2, y_2), and (x_3, y_3) be three points such that $x_1 \neq x_2$ and $x_1 \neq x_3$. If the points cannot be renumbered to satisfy this condition, the points cannot be on a circle. Let the center of the circle be (x_0, y_0) and the radius of the circle be r. Then

$$y_0 = \frac{k_2 - k_1}{n_2 - n_1}$$

$$x_0 = k_2 - n_2 y_0$$

and

$$r = \sqrt{(x_1 - x_0)^2 + (y_1 - y_0)^2}$$

where

$$k_1 = \frac{1}{2}[(x_1 + x_2) + n_1(y_1 + y_2)]$$

$$k_2 = \frac{1}{2}[(x_1 + x_3) + n_2(y_1 + y_3)]$$

$$n_1 = \frac{y_1 - y_2}{x_1 - x_2}$$

and

$$n_2 = \frac{y_1 - y_3}{x_1 - x_3}$$

If $n_1 = n_2$, the points cannot form a circle.

Examples:

1. Find the equation of the circle that goes through the three points (1, 1), (3.5, −7.6), and (12, 0.8).
2. Find the equation of the circle that passes through the three points (0, 1), (−1, 0), and (0, 1).

Solutions:

1. $n_1 = -3.44$, $k_1 = 13.60$, $n_2 = -.02$, $k_2 = 6.48$
   Center = (6.45, −2.08), $r = 6.26$
   Equation: $(x - 6.45)^2 + (y + 2.08)^2 = (6.26)^2$
2. $n_1 = 1.00$, $k_1 = 0.00$, $n_2 = -1.00$, $k_2 = 0.00$
   Center = (0, 0), $r = 1$
   Equation: $x^2 + y^2 = 1$
   Note: (−1, 0) must be (x_1, y_1)

COMPOUNDED AMOUNT

Let

n = number of time periods
i = periodic interest rate expressed as a decimal, e.g., 6% is represented as .06
PV = present value or principal
FV = future value or amount
I = interest amount

Each value can be calculated from the others by the following formulas:

1. $FV = PV(1 + i)^n$

2. $PV = FV(1 + i)^{-n}$

3. $n = \dfrac{\ln (FV/PV)}{\ln (1 + i)}$

4. $i = \left(\dfrac{FV}{PV}\right)^{1/n} - 1$

5. $I = PV[(1 + i)^n - 1]$

DEPRECIATION SCHEDULES (STRAIGHT LINE)

Let

PV = original value of asset (less salvage value)
n = lifetime number of periods of asset
B_k = book value at time period K
D = each year's depreciation
k = number of time period, i.e., 1, 2, 3, . . ., or n

Then, B_k and D can be calculated by the following formulas:

1. $D = PV/n$

2. $B_k = PV - kD$

DEPRECIATION SCHEDULES (SUM-OF-THE-YEAR'S DIGITS)

Let

n = lifetime number of periods of asset
S = salvage value
D_k = depreciation over time period k
B_k = book value at time period k

PV = original value of asset (less salvage value)
  k = number of time period, i.e., 1, 2, 3, . . ., or n

Then, D_k and B_k can be calculated by the following formulas:

1. $D_k = \dfrac{2(n - k + 1)}{n(n + 1)} \, PV$

2. $B_k = S + \dfrac{(n - k)D_k}{2}$

DEPRECIATION SCHEDULES
(VARIABLE RATE DECLINING BALANCE)

Let

PV = original value of asset (less salvage value)
  n = lifetime periods of asset
  R = depreciation rate (given by user)
  D_k = depreciation at time period k
  B_k = book value at time period k
  k = number of time period, i.e., 1, 2, 3, . . ., or n

Then, D_k and B_k can be calculated by the following formulas:

1. $D_k = PV \dfrac{R}{n} \left(1 - \dfrac{R}{n} \right)^{k-1}$

2. $B_k = PV \left(1 - \dfrac{R}{n} \right)^{k}$

DETERMINANT AND INVERSE OF A 2 × 2 MATRIX

Let

$$A = \begin{bmatrix} u_{11} & u_{12} \\ a_{21} & a_{22} \end{bmatrix} \text{ be a 2 × 2 matrix}$$

The determinant of A denoted by Det A or $|A|$ is evaluated by the following formula:

$$\text{Det } A = a_{22}\, a_{11} - a_{12}\, a_{21}$$

The multiplicative inverse A^{-1} of A can be determined from:

$$A^{-1} = \begin{bmatrix} a_{22}/\text{Det } A & -a_{12}/\text{Det } A \\ -a_{21}/\text{Det } A & a_{11}/\text{Det } A \end{bmatrix}$$

DETERMINANT OF A 3 × 3 MATRIX

Let

$$A = \begin{bmatrix} a_{11} & a_{12} & a_{13} \\ a_{21} & a_{22} & a_{23} \\ a_{31} & a_{32} & a_{33} \end{bmatrix} \text{ be a 3 × 3 matrix}$$

The determinant of A denoted by Det A or $|A|$, is calculated by expanding A by minors about the first column. The formula is:

$$\text{Det } A = a_{11}\begin{vmatrix} a_{22} & a_{23} \\ a_{32} & a_{33} \end{vmatrix} - a_{21}\begin{vmatrix} a_{12} & a_{13} \\ a_{32} & a_{33} \end{vmatrix} + a_{31}\begin{vmatrix} a_{12} & a_{13} \\ a_{22} & a_{23} \end{vmatrix}$$

$$= a_{11}[a_{22}\, a_{33} - a_{23}\, a_{32}] - a_{21}\,[a_{33}\, a_{12} - a_{32}\, a_{13}]$$

$$+ a_{31}\,[a_{23}\, a_{12} - a_{13}\, a_{22}]$$

Example:

Find the determinant of

$$A = \begin{bmatrix} -1 & 0 & 3 \\ 7 & 1 & -1 \\ 2 & 3 & 0 \end{bmatrix}$$

Solution:

$$\text{Det } A = 54$$

DIRECT REDUCTION LOAN INTEREST RATE

These formulas compute the interest rate on a mortgage where payments are made at the end of the period. Let

n = number of payments
i = periodic interest rate expressed as a decimal, e.g., 6% is represented as .06
PMT = payment
PV = present value or principal

The equation $f(i)$ can be solved by an iteration for i using Newton's method:

$$i_{k+1} = i_k - \frac{f(i)}{f'(i)}$$

where

$$f(i) = \frac{1 - (1 + i)^{-n}}{i} - \frac{PV}{PMT}$$

and

$$f'(i) \text{ is the first derivative of } f(i)$$

An initial guess of i_0 must be made.

Example:

Compute the monthly interest rate on a mortgage of $30,000. The loan requires 360 monthly payments of $179.86 to be payed off.

Solution: 0.5%

DIRECT REDUCTION LOAN PAYMENT, PRESENT VALUE, NUMBER OF TIME PERIODS

The following formulas determine the accumulated interest and remaining balance of a mortgage. Let

n = number of payment periods
PV = present value or principal

PMT = payment
   i = periodic interest rate expressed as a decimal

Then, PMT, PV, and n can be calculated from the other three by the following formulas:

1. $PMT = PV \left[\dfrac{i}{1 - (1 + i)^{-n}} \right]$

2. $PV = PMT \left[\dfrac{1 - (1 + i)^{-n}}{i} \right]$

3. $n = - \dfrac{\ln (1 - iPV/PMT)}{\ln (1 + i)}$

DIRECT REDUCTION LOAN ACCUMULATED INTEREST, REMAINING BALANCE

The following formulas determine the accumulated interest and remaining balance of a mortgage. Let

I_{c-k} = the accumulated interest paid by payments c through k
PV_k = the remaining balance after payment k
   n = number of payments
   i = periodic interest rate expressed as a decimal, e.g., 6% is expressed
       as .06
   $j = c - 1$

Then, I_{c-k} and PV_k can be calculated by the following formulas:

1. $I_{c-k} = PMT \left\{ k - j - \dfrac{(1 + i)^{k-n}}{i} [1 - (1 + i)^{j-k}] \right\}$

2. $PV_k = \dfrac{PMT}{i} [1 - (1 + i)^{k-n}]$

DIRECT REDUCTION LOAN AMORTIZATION SCHEDULES

These formulas compute a table of interest paid, payment to principal, and present value of a mortgage. They also can be used to find accumulated interest. Let

I_k = interest paid in kth payment
PMT = payment
PP_k = payment to principal of kth payment
PV_k = remaining balance after kth payment
PV_0 = amount of loan
i = periodic interest rate expressed as a decimal, e.g., 6% is expressed as .06

An amortization schedule consists of the interest paid, the payment to principal, and the remaining balance for each payment k (where $k = 1, 2, \ldots$).

These quantities are calculated by the following formulas:

1. $I_k = i\, PV_{k-1}$

2. $PP_k = PMT - I_k$

3. $PV_k = PV_{k-1} - PP_k$

DISCOUNTED CASH FLOW ANALYSIS

Let

PV_0 = original investment
PV_k = cash flow of kth period
i = discount rate per period as a decimal, e.g., 6% is expressed as .06
C_k = net present value at period k

Then

$$C_k = -PV_0 + \sum_{k=1}^{n} \frac{PV_k}{(1 + i)^k}$$

Example:

You are offered an investment opportunity for $100,000 at a capital cost of 10%

after taxes. Will this investment be profitable based on the following cash flows?

Year	Cash Flow
1	$34,000
2	$27,500
3	$59,700
4	$ 7,800

Solution:

$$C_1 = \$-69,090.91$$
$$C_2 = \$-46,363.64$$
$$C_3 = \$-1,510.14$$
$$C_4 = \$3817.36$$

Since C_4 is positive the cash flow is profitable to the extent that the cost of capital is 10%.

GEOMETRIC RELATIONS

BASIC DEFINITIONS IN PLANE GEOMETRY

FORMULAS OF STRAIGHT LINES:
(refer to Figure A4)

Line parallel to y-axis

$$x = a$$

Line parallel to x-axis

$$y = b$$

General equation of straight line

$$y = mx + b$$

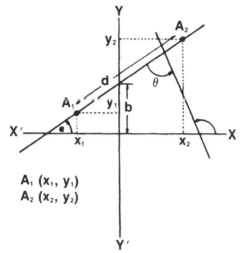

Figure A4. Lines plotted on rectangular coordinates.

Line through one point

$$y - y_1 = m(x - x_1)$$

Distance between two points A_1 and A_2

$$d = \sqrt{(x_2 - x_1)^2 + (y_2 - y_1)^2}$$

Point of intersection of two straight lines

$$x' = \frac{b_2 - b_1}{m_1 - m_2}$$

Point dividing $A_1 A_2$ in ratio r/s

$$\left(\frac{rx_2 + sx_1}{r + s}, \frac{ry_2 + sy_1}{r + s} \right)$$

Midpoint of $A_1 A_2$

$$\left(\frac{x_1 + x_2}{2}, \frac{y_1 + y_2}{2} \right)$$

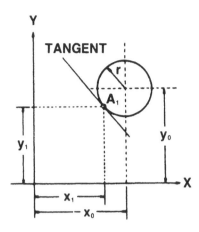

Figure A5. Circle and parameters of interest.

Slope of A_1 A_2

$$m = \tan \alpha = \frac{y_2 - y_1}{x_2 - x_1}$$

Angle θ between two lines of slopes m_1 and m_2

$$\tan \theta = \frac{m_2 - m_1}{1 + m_1 m_2}$$

EQUATIONS FOR A CIRCLE
(refer to Figure A5)

Center at the origin, radius r

$$x^2 + y^2 = r^2$$

Center at (h,k), radius r

$$(x - h)^2 + (y - k)^2 = r^2$$

Radius of circle

$$r = \sqrt{x_0^2 + y_0^2 - c}$$

Tangent at point $A_1(x_1, y_1)$

$$y = \frac{r^2 - (x - x_0)(x_1 - x_0)}{y_1 - y_0} + y_0$$

EQUATIONS FOR AN ELLIPSE
(refer to Figure A6)

Eccentricity

$$e = \frac{\sqrt{a^2 - b^2}}{a}$$

Distance from center to either focus

$$\sqrt{a^2 - b^2}$$

Sum of distances from any point on ellipse to foci

$$2a$$

Center at origin with foci on $X'X$

$$\frac{x^2}{a^2} + \frac{y^2}{b^2} = 1$$

Center at origin with foci on $Y'Y$

$$\frac{x^2}{b^2} + \frac{y^2}{a^2} = 1$$

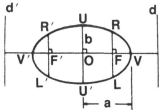

O—CENTER/V,V′—VERTICES/V′V—MAJOR AXIS = 2a
U′U—MINOR AXIS = 2b/F,F′—FOCI/d,d′—DIRECTRICES
LR,L′R′—LATERA RECTA

Figure A6. Ellipse, where major axis = $2a$; minor axis = $2b$; eccentricity = e.

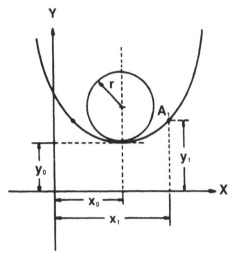

Figure A7. Details of the parabola.

Center at point (h,k), major axis parallel to $X'X$

$$\frac{(x - h)^2}{a^2} + \frac{(y - k)^2}{b^2} = 1$$

Center at point (h,k), major axis parallel to $Y'Y$

$$\frac{(x - h)^2}{b^2} + \frac{(y - k)^2}{a^2} = 1$$

EQUATIONS FOR A PARABOLA
(refer to Figure A7)

Note

p = distance from the vertex to the focus
e = eccentricity

Parabola open at top

$$x^2 = 2py \text{ (at the origin)}$$

$$(x - x_0)^2 = 2p(y - y_0) \text{ (elsewhere)}$$

Parabola open at bottom

$$x^2 = -2py \text{ (at the origin)}$$

$$(x - x_0)^2 = -2p(y - y_0) \text{ (elsewhere)}$$

General formula

$$y = ax^2 + bx + c$$

Tangent at point $A_1(x_1, y_1)$

$$y = \frac{2(y_1 - y_0)(x - x_1)}{x_1 - x_0} + y_1$$

Vertex radius

$$r = p$$

EQUATIONS FOR A HYPERBOLA
(refer to Figure A8)

General formula

$$ax^2 + by^2 + cx + dy + e = 0$$

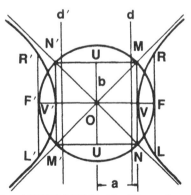

O—CENTER/V,V′—VERTICES/V′V—TRANSVERSE AXIS = 2a
U′U—CONJUGATE AXIS = 2b/F,F′—FOCI
d,d′—DIRECTRICES/LR, L′R′—LATERA RECTA
M′M & N′N LINES—ASYMPTOTES

Figure A8. Illustrates a hyperbola ($e > 1$).

Eccentricity

$$e = \sqrt{a^2 + b^2}$$

Gradient of asymptotes

$$\tan \alpha = m = \pm \frac{b}{a}$$

Vertex radius

$$p = \frac{b^2}{a}$$

Center at origin, foci on $X'X$

$$\frac{x^2}{a^2} - \frac{y^2}{b^2} = 1$$

Slopes of asymptotes

$$\pm b/a$$

Center at origin, foci on $Y'Y$

$$\frac{y^2}{a^2} - \frac{x^2}{b^2} = 1$$

Slopes of asymptotes

$$\pm a/b$$

Center at (h,k), transverse axis parallel to $X'X$

$$\frac{(x - h)^2}{a^2} - \frac{(y - k)^2}{b^2} = 1$$

Slopes of asymptotes

$$\pm b/a$$

Center at (h,k), transverse axis parallel to $Y'Y$

$$\frac{(y - k)^2}{a^2} - \frac{(x - h)^2}{b^2} = 1$$

Slopes of asymptotes

$$\pm a/b$$

Center at origin, $X'X$ and $Y'Y$ for asymptotes

$$xy = c$$

Center at (h,k), asymptotes parallel to $X'X$ and $Y'Y$

$$(x - h)(y - k) = c$$

For a rectangular hyperbola:

$$a = b$$
$$e = \sqrt{2}$$
asymptotes are perpendicular

FORMULAS FOR AREAS

Configuration	Formula	Figure Reference
Circle	$A = (1/4)\pi d^2$	A9(a)
	$C = 2\pi r = \pi d$	
Square	$A = a^2$	A9(b)
	$a = \sqrt{A}$, $d = a\sqrt{2}$	
Rectangle	$A = ab$	A9(c)
	$d = \sqrt{a^2 + b^2}$	

(continued)

Configuration	Formula	Figure Reference
Parallelogram	$A = ah = ab \sin \alpha$	A9(d)
	$d_1 = \sqrt{(a + h \cot \alpha)^2 + h^2}$	
	$d_2 = \sqrt{(a - h \cot \alpha)^2 + h^2}$	

FORMULAS FOR AREAS

Configuration	Formula	Figure Reference
Trapezoid	$A = \dfrac{a + b}{2} h = mh$	A10(a)
	$m = \dfrac{a + b}{2}$	
Triangle	$A = \dfrac{ah}{2} = qs$	A10(b)
	$= \sqrt{s(s - a)(s - b)(s - c)}$	
	$s = \dfrac{a + b + c}{2}$	
Equilateral Triangle	$A = \dfrac{a^2}{4} \sqrt{3}$	A10(c)
	$h = \dfrac{a}{2} \sqrt{3}$	
Pentagon	$A = (5/8) \, r^2 \sqrt{10 + 2\sqrt{5}}$	A10(d)
	$a = (1/2) \, r \sqrt{10 - 2\sqrt{5}}$	
	$q = (1/4) \, r \sqrt{6 + 2\sqrt{5}}$	

(continued)

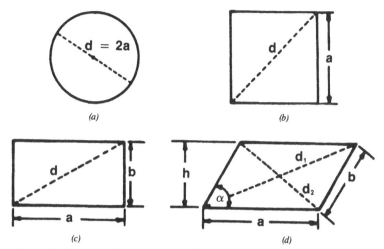

Figure A9. Geometric figures: (a) circle; (b) square; (c) rectangle; (d) parallelogram.

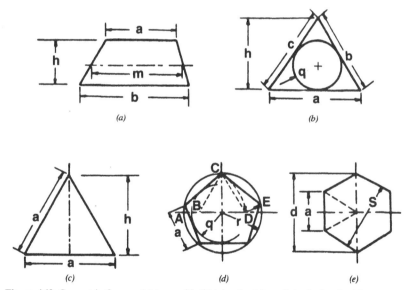

Figure A10. Geometric figures: (a) trapezoid; (b) triangle; (c) equilateral triangle; (d) pentagon; (e) hexagon.

Configuration	Formula	Figure Reference
Hexagon	$A = \dfrac{3a^2 \sqrt{3}}{2}$	A10(e)
	$d = 2a$	
	$\quad = 1.155\ s$	
	$s = 0.866\ d$	

LOCATION OF CENTROIDS FOR VARIOUS GEOMETRIES

Location of centroids for various geometries.

Plane Geometry	Centroid Location
Perimeter of triangle	Center of inscribed circle of the triangle whose vertices are the midpoints of the sides of the given triangle
Arc of semicircle of radius R	Distance from diameter $= \dfrac{2R}{\pi}$
Area of 2α radians of a circle of radius R	Distance from center of circle $= \dfrac{R \sin \alpha}{\alpha}$
Area of triangle	Intersection of the medians
Area of quadrilateral	Intersection of the diagonals of the parallelogram whose sides pass through adjacent trisection points of pairs of consecutive sides of the quadrilateral
Area of semicircle of radius R	Distance from diameter $= \dfrac{4R}{3\pi}$
Area of circular sector of radius R and central angle 2α radians	Distance from center of circle $= \dfrac{2R \sin \alpha}{3\alpha}$

Location of centroids for various geometries *(continued)*.

Plane Geometry	Centroid Location
Area of semiellipse of altitude h	Distance from base $= \dfrac{4h}{3\pi}$
Area of a quadrant of an ellipse of major and minor semiaxes a and b	Distance from minor axis $= \dfrac{4a}{3\pi}$, distance from major axis $= \dfrac{4b}{3\pi}$
Area of right parabolic segment of altitude h	Distance from base $= (2/5)h$

Solid Geometry	Centroid Location
Lateral area of regular pyramid or right circular cone	Distance from base $= (1/3)h$
Area of hemisphere *of radius R*	Distance from base $= (1/2)R$
Volume of pyramid or cone	One fourth way from the centroid of the base to the vertex of the pyramid or cone

FORMULAS FOR SOLID BODIES

Object	Formula	Figure Reference
Cube	$V = a^3$	A11(a)
	$A_0 = 6a^2$	
	$d = a\sqrt{3}$	
Cuboid	$V = abc$	A11(b)

(continued)

Object	Formula	Figure Reference
	$A_0 = 2(ab + ac + bc)$	
	$d = \sqrt{a^2 + b^2 + c^2}$	
Pyramid	$V = A_1 h/3$	A11(c)
Cylinder	$V = \dfrac{d^2 \pi}{4} h$	A11(d)
	$A_m = 2\pi r h$	
	$A_0 = 2\pi r(r + h)$	

V = volume; A_0 = surface area.

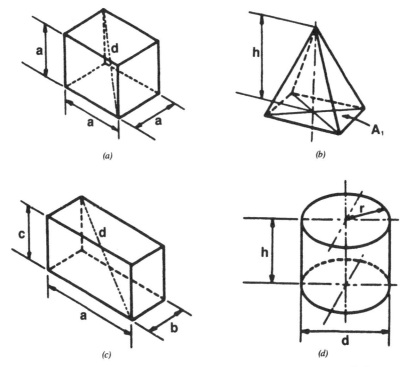

(a)   (b)

(c)   (d)

Figure A11. Geometric figures: (a) cube; (b) pyramid; (c) cuboid; (d) cylinder.

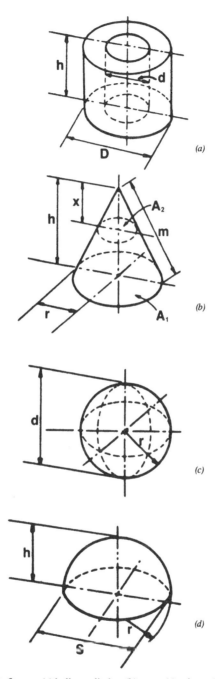

Figure A12. Geometric figures: (a) hollow cylinder; (b) cone; (c) sphere; (d) segment of a sphere.

FORMULAS FOR SOLID BODIES

Object	Formula	Figure Reference
Hollow cylinder	$V = \dfrac{h\pi}{4}\,(D^2 - d^2)$	A12(a)
Cone	$V = \dfrac{r^2 \pi h}{3}$	A12(b)
	$A_m = r\pi m$	
	$A_0 = r\pi(r + m)$	
	$m = \sqrt{h^2 + r^2}$	
	$A_2/A_1 = x^2/h^2$	
Sphere	$V = (4/3)\,\pi r^3 = (1/6)\,\pi d^3$	A12(c)
	$A_0 = 4\pi r^2 = \pi d^2$	
Segment of a sphere	$V = \dfrac{\pi h}{6}\left(\dfrac{3}{4}\,s^2 + h^2\right)$	A12(d)
	$= \pi h^2\left(r - \dfrac{h}{3}\right)$	
	$A_m = 2\pi rh$	
	$= \dfrac{\pi}{4}\,(s^2 + 4h^2)$	

V = volume; A_0 = surface area.

FORMULAS FOR SOLID BODIES

Object	Formula	Figure Reference
Sliced cylinder	$V = \dfrac{d^2 \pi}{4}\,h$	A13(a)

(continued)

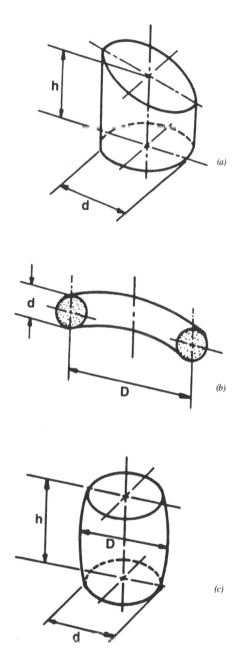

Figure A13. Geometric figures: (a) sliced cylinder; (b) torus; (c) barrel.

Object	Formula	Figure Reference
Torus	$V = \dfrac{D\pi^2 d^2}{4}$	A13(b)
	$A_0 = Dd\pi^2$	
Barrel	$V = \dfrac{h\pi}{12}(2D^2 + d^2)$	A13(c)

V = volume; A_0 = surface area.

INVERSE HYPERBOLIC FUNCTIONS

Inverse hyperbolic functions are determined by the following formulas:

1. $\sinh^{-1} x = \ln\left[x + (x^2 + 1)^{1/2}\right]$

2. $\cosh^{-1} x = \ln\left[x + (x^2 - 1)^{1/2}\right]$ $x \geq 1$

3. $\tanh^{-1} x = \dfrac{1}{2}\ln\left[\dfrac{1 + x}{1 - x}\right]$ $x^2 < 1$

4. $\mathrm{csch}^{-1} x = \sinh^{-1}\left[\dfrac{1}{x}\right]$ $x \neq 0$

5. $\mathrm{sech}^{-1} x = \cosh^{-1}\left[\dfrac{1}{x}\right]$ $0 < x \leq 1$

6. $\coth^{-1} x = \tanh^{-1}\left[\dfrac{1}{x}\right]$ $x^2 > 1$

LINEAR INTERPOLATION

If $(x_1, f(x_1))$ and $(x_2, f(x_2))$ are two points of a function $f(x)$, then the function at x_0 can be approximated by the following formula:

$$f(x_0) \cong \frac{(x_2 - x_0)\,f(x_1) + (x_0 - x_1)\,f(x_2)}{(x_2 - x_1)}$$

This is called the linear interpolation formula. Of course, x_2 cannot equal x_1.

Example:

If (1.2, .30119) and (1.3, .27253) are two points of a function, find $f(1.27)$ and $f(1.29)$.

Solution:

1. $f(1.27) = .28113$
2. $f(1.29) = .27540$

MATRIX MULTIPLICATION (2 × 2)

Let

$$A = \begin{bmatrix} a_{11} & a_{12} \\ a_{21} & a_{22} \end{bmatrix} \quad \text{and} \quad B = \begin{bmatrix} b_{11} & b_{12} \\ b_{21} & b_{22} \end{bmatrix}$$

be two 2 × 2 matrices. The matrix product of A and B is calculated as follows:

$$AB = \begin{bmatrix} a_{11}b_{11} + a_{12}b_{21} & a_{11}b_{12} + a_{12}b_{22} \\ a_{21}b_{11} + a_{22}b_{21} & a_{21}b_{12} + a_{22}b_{22} \end{bmatrix}$$

Let the answer be denoted by:

$$C = \begin{bmatrix} c_{11} & c_{12} \\ c_{21} & c_{22} \end{bmatrix}$$

Example:

Find the product of the two matrices:

$$A = \begin{bmatrix} 1 & 2 \\ -1 & 3 \end{bmatrix} \quad \text{and} \quad B = \begin{bmatrix} 1 & -1 \\ 2 & 4 \end{bmatrix}$$

Solution:

$$C = \begin{bmatrix} 5 & 7 \\ 5 & 13 \end{bmatrix}$$

MATRIX INVERSION

OF A 3 × 3 MATRIX

If a_{ij} indicates a number in the ith row, jth column then a 3 × 3 matrix A can be represented as

$$\begin{bmatrix} a_{11} & a_{12} & a_{13} \\ a_{21} & a_{22} & a_{23} \\ a_{31} & a_{32} & a_{33} \end{bmatrix}$$

then the multiplicative inverse of A is denoted by A^{-1} and is calculated as follows:

$$A^{-1} = \begin{bmatrix} \dfrac{\begin{vmatrix} a_{22} & a_{23} \\ a_{32} & a_{33} \end{vmatrix}}{\text{Det } A} & -\dfrac{\begin{vmatrix} a_{12} & a_{13} \\ a_{32} & a_{33} \end{vmatrix}}{\text{Det } A} & \dfrac{\begin{vmatrix} a_{12} & a_{13} \\ a_{22} & a_{23} \end{vmatrix}}{\text{Det } A} \\[3em] -\dfrac{\begin{vmatrix} a_{21} & a_{23} \\ a_{31} & a_{33} \end{vmatrix}}{\text{Det } A} & \dfrac{\begin{vmatrix} a_{11} & a_{13} \\ a_{31} & a_{33} \end{vmatrix}}{\text{Det } A} & -\dfrac{\begin{vmatrix} a_{11} & a_{13} \\ a_{21} & a_{23} \end{vmatrix}}{\text{Det } A} \\[3em] \dfrac{\begin{vmatrix} a_{21} & a_{22} \\ a_{31} & a_{32} \end{vmatrix}}{\text{Det } A} & -\dfrac{\begin{vmatrix} a_{11} & a_{12} \\ a_{31} & a_{32} \end{vmatrix}}{\text{Det } A} & \dfrac{\begin{vmatrix} a_{11} & a_{12} \\ a_{21} & a_{22} \end{vmatrix}}{\text{Det } A} \end{bmatrix}$$

For the ith, jth position of A^{-1} use the minor of the jth, ith position of the original matrix. The minor is the 2 × 2 matrix left after crossing out the ith row and jth column of A.

NOMOGRAPH CONSTRUCTION[1]

Nomographs or alignment charts are used to solve problems graphically. Although the use of computers has largely replaced graphical methods, nomo-

[1]Suggested reference—P. Carroll, *How to Chart Data*, McGraw Hill Book Co., New York (1960).

graphs are highly useful in presentations and for quick in-the-field calculations. Simple procedures for constructing nomographs are outlined below.

SIMPLE ADDITION

Figure A14 illustrates the procedure described below.

Step 1. To prepare a nomograph for simple addition, start with a squared

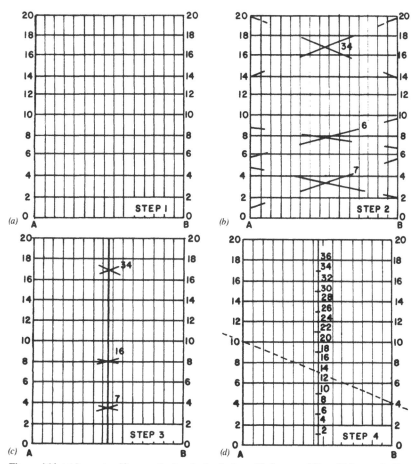

Figure A14. (a) Lay out uniform vertical scales beginning with the same 0 line. (b) Locate the line of the sums. At the bottom, 6 + 1 equals 2 + 5. Near the center, 6 + 10 equals 7 + 9. At the top, 20 + 14 equals 14 + 20. Only the intersections need be drawn. (c) Three or more intersections should line up vertically. When they do, they locate the straight line of the sums. Draw in the line. (d) With the sum line located you may erase the intersection marks. Then you can lay out the sum scale.

sheet of paper, preferably with graph line divisions. Along both sides of page lay out vertical scales in a uniform fashion (i.e., 0, 1, 2, 3, 4, . . ., 10).

Step 2. Locate the line of SUMS. Connect any numbers on the *A* scale with any number on the *B* scale. Draw a short line near the center. Rotate the straight edge to connect any other two numbers having the same sum and draw a short line to mark the intersection with the first line. Note the sum along the short line just drawn. Repeat the process for at least two more sums.

Step 3. Draw a straight line through the intersections of the 3 sum lines and erase intersecting lines.

Step 4. Layout the sum scale in the same manner for as many points as desired. Any easy approach is to simply lay out sums horizontally on each cross line. Note that the two scales are equal, the sum scale is simply twice either scale.

SUBTRACTION

To prepare a subtraction chart, reverse direction in the layout. So for example, instead of $4 + 10 = 14$ in Figure A14(d), use $14 - 10 = 4$ on the middle scale.

DIVISION

This technique uses simple proportion graphically. Either uniform or logarithmic scales can be used. Figure A15 illustrates a uniform scale projection. The scale is turned at an angle which brings the number of divisions counted off on the scale within the range of the length of line you want to divide. Mark off lightly the divisions [refer to Figure A15(b)].

If the preferred scale is logarithmic, use a cross section of log scale paper as shown in Figure A16 (i.e., use semi-logarithmic paper instead of the ruler).

MULTIPLICATION

Figures A16 and A17 illustrate construction of this type of nomograph. Rotate a piece of semi-logarithmic paper to an angle where the tenth division exactly coincides with a line connecting both 10s, while the 1 on the logarithmic paper rests on a line connecting the 1s of the scales. This makes one layout of points correct for both scales. Next mark off the main divisions. The dots in the figure are lined up on the angle. You should mark off all divisions in constructing a chart for problem solving. Once the divisions are marked, set the triangles so as to get a right angle with the vertical scales and lay out both scales from one set of points.

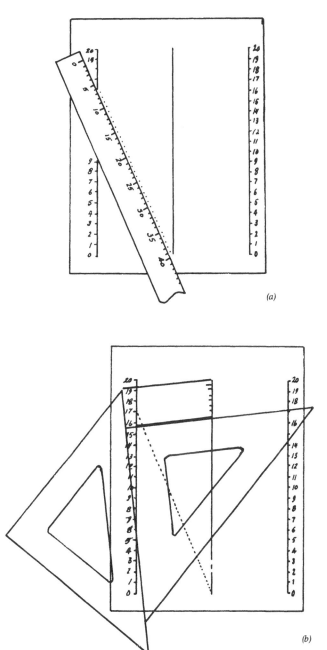

(a)

(b)

Figure A15. (a) Turn a convenient scale to an angle sufficient to allow the required number of divisions to be marked off. (b) Set up triangles so as to connect the last mark with the corresponding division on your scale. Then, holding the parallel, mark off all the divisions.

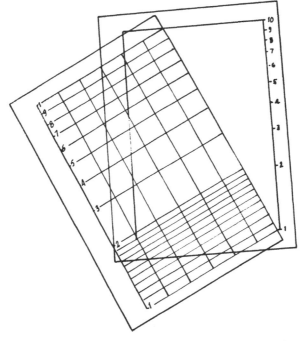

Figure A16. To get logarithmic divisions, you can use cross-section paper with the desired scale. You will find it convenient to use if you will fold or cut the sheet to bring the scale to the edge of the paper.

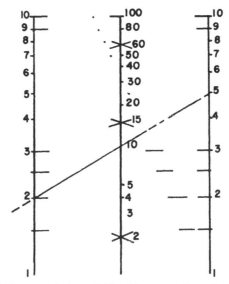

Figure A17. You can multiply or divide with nomographs made with log scales.

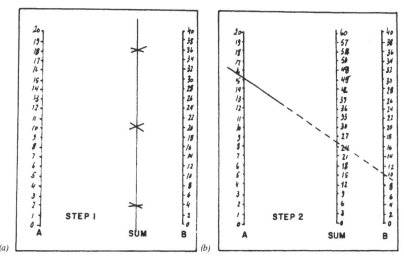

Figure A18. Nomographs may be used to add two variables having unlike ranges. (a) The Sum line is located by intersections. It falls one-third of the distance from the double scale. (b) The Sum Scale is graduated to provide answers to our problems of adding two numbers.

Since the scales are logarithmic, you can use them to multiply. In reality, this nomograph works in the identical fashion as the addition example, except that we are adding logarithms (hence, multiplying).

UNEQUAL SCALES

The procedure is outlined below for addition, however, the approach for multiplying by substituting log scales for uniform scales is the same. Figure A18 illustrates a case where one scale is twice the other. The procedure for preparing the nomograph is identical to that outlined under Simple Addition. By drawing three sets of intersections, we determine the location of the Sum Scale. This scale will be closer to the scale having the larger values per unit.

As outlined earlier, proceed with laying out the Sum Scale (refer to Figure A18). Write in the sums of the numbers that are horizontally opposite. This example results in a scale based on 35, i.e., 0, 3, 6, 9, . . . 60. One can also set the scale of the divisions at an angle as shown earlier. Then mark off the Sum Scale by proportion.

MORE THAN TWO VARIABLES

The construction procedure for three variables is illustrated in Figure A19. Continuing with the above example, we lay out a third scale on line *A* (we use

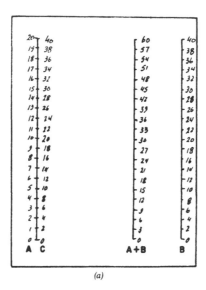

(a)

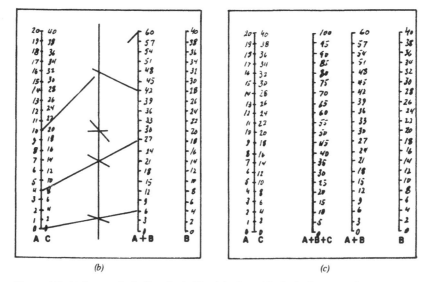

(b)

(c)

Figure A.19. (a) Lay out Scale C on the inside of the line of Scale A. (b) Locate the new line of sums adding A + B to C. (c) Mark off the new scale showing the sum of A + B + C.

an existing scale to achieve maximum accuracy in reading the chart). We refer to this new scale as Scale *C*. Next find a new Sum Scale just as before. Using trial sums, make several intersections to locate the Sum Scale for *A* + *B* + *C* [refer to Figure A19(b)]. The location is established by the ratio of the Sum *A* + *B* scale of the three values per unit to the *C* scale of the two values per unit.

Having constructed the final line, lay out its proper scale as shown in Figure A19(c). The same three methods previously described for graduating can be used.

The procedure for laying out four variables is shown in Figure A20. Again, continuing with the same example, locate the fourth scale on an available line. The one farthest away is the line of Scale *B*. Lay out the fourth scale alongside Scale *B* and name it Scale *D* [refer to Figure A20(b)].

Proceed to locate the new line of sums. Ignore all the other lines except the Sum (*A* + *B* + *C*) and the new Scale *D*. Construct several intersections to locate the scale [refer to Figure A20(c)]. Once located, lay out the final scale of sums.

You can erase the graduations on the sum lines for (*A* + *B*) and (*A* + *B* + *C*) since they are only intermediate answers (i.e., construction scales). The final chart is shown in Figure A20(d). An example is illustrated on the chart to show how to use it.

POLYNOMIAL EVALUATION

A polynomial of the form

$$f(x) = a_0 x^n + a_1 x^{n-1} + \ldots + a_{n-1} x + a_n$$

is evaluated by writing it in the form

$$x (\ldots x(x(a_0 x + a_1) + a_2) + \ldots + a_{n-1}) + a_n$$

n can be any positive integer. Refer to the statistics section for further notes.

POLYGONS CIRCUMSCRIBED ABOUT A CIRCLE

Given the radius of a circle one can calculate the length S_2 of a side and the area A_2 of a polygon of *n* sides circumscribed about a circle. The formulas used are:

1. $S_2 = 2r \tan (c/n)$

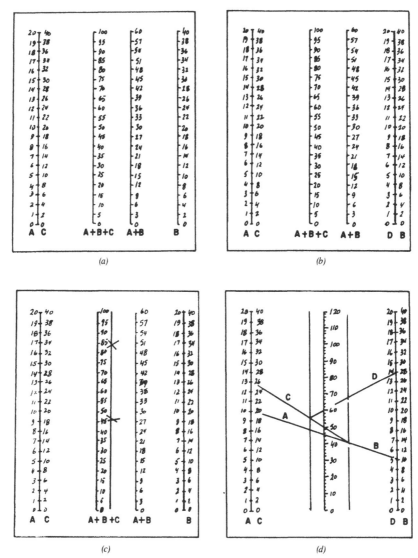

Figure A20. Four variables are combined in a nomograph as shown by the example $A = 10$, $B = 10$, $C = 26$, $D = 14$ and the total is 60. (a) We begin the addition of a fourth variable with a nomograph that combines three already located. (b) The fourth scale D is added. Usually, this is laid out on a line available farthest away from the sum to be added to it. (c) The line of final Sum is located by intersections as described earlier in this chapter. Then it is graduated to provide the final Sums. (d) With all lines located and scales laid out, we can erase construction details. The Sum lines for $A + B$ and $A + B + C$ become reference lines.

2. $A_2 = n\, r^2 \tan (c/n)$

where

$c = 2 \sin^{-1} 1 = \pi$ radians $= 180° = 200$ grads
$n =$ number of sides
$r =$ radius of the circle

It must be true that n is an integer greater than 2.

Example:

Find the length of a side and the area of a polygon of 6 sides circumscribed about a circle of radius 5.

Solution:

$$S_2 = 5.77$$
$$A_2 = 86.60$$

POLYGONS INSCRIBED IN A CIRCLE

Given the radius of a circle one can calculate the length S_1 of a side and the area A_1 of a polygon of n sides inscribed in the circle. The formulas used are:

1. $S_1 = 2r \sin \left(\dfrac{c}{n}\right)$

2. $A_1 = \dfrac{1}{2} n\, r^2 \sin \left(\dfrac{2c}{n}\right)$

where

$c = 2 \sin^{-1} 1 = \pi$ radians $= 180° = 200$ grads
$n =$ number of sides
$r -$ radius of the circle

It must be true that n is an integer greater than 2.

Example:

Given a circle of radius 5 find the length of a side and area of a polygon of 6 sides inscribed in the circle.

Solution:

$$S_1 = 5.00$$
$$A_1 = 64.95$$

QUADRATIC EQUATIONS

A general quadratic equation is of the form:

$$ax^2 + bx + c = 0$$

The equation has two roots, x_1 and x_2. Let

$$D = \frac{b^2 - 4ac}{4a^2}$$

If $D \geq 0$, then

$$x_1 = \begin{cases} -\dfrac{b}{2a} + \sqrt{D} & \text{if } -\dfrac{b}{2a} \geq 0 \\[2em] -\dfrac{b}{2a} - \sqrt{D} & \text{if } -\dfrac{b}{2a} < 0 \end{cases}$$

and

$$x_2 = \frac{c}{ax_1}$$

These formulas compute the larger root (in absolute value) first. Better significance can be obtained by this method.

If $D < 0$, then

$$x_1, x_2 = -\frac{b}{2a} \pm \sqrt{-D} = u \pm iv$$

The coefficient a cannot be zero.

Examples:

Find the solution to the following three equations:

1. $x^2 - 3x - 4 = 0$
2. $2x^2 + 5x + 3 = 0$
3. $2x^2 + 3x + 4 = 0$

Solution:

1. $D = 6.25$ $x_1 = 4, x_2 = -1$
2. $D = .06$ $x_1 = -1.50, x_2 = -1.00$
3. $D = -1.44$ $x_1, x_2 = -.75 \pm 1.20i$

SERIES FORMULAS[2]

ARITHMETIC SERIES

The series 1, 4, 7, 10, . . . is an arithmetic series. The difference between two consecutive terms is constant.

$$a_n = a_0 + (n - 1) \Delta a$$

$$\Sigma a_n = \frac{n}{2} (a_0 + a_n) = a_0 n + \frac{n (n - 1) \Delta a}{2}$$

where

a_0 = initial term
a_n = final term
n = number of terms

For the arithmetic mean, each term in the arithmetic series is the arithmetic mean of its adjacent terms:

$$x = (a_1 + a_2)/2$$

[2]Suggested reference—Jaquez, J. A., *A First Course in Computing and Numerical Methods*, Addison-Wesley Publishing Co., Reading, Mass (1970).

GEOMETRIC SERIES

$$a_n = a_0 q^{n-1}$$

$$\Sigma a_n = a_0 \frac{q^n - 1}{q - 1} = \frac{q\, a_n - a_0}{q - 1}$$

where q = quotient of two consecutive terms.

Each term in a geometric series is the geometric mean of its adjacent terms:

$$x = \sqrt{a_1 a_2}$$

BINOMIAL SERIES

$$f(x) = (1 \pm x)^\alpha = 1 \pm \binom{\alpha}{1} x + \binom{\alpha}{2} x^2 \pm \binom{\alpha}{3} x^3 + \ldots$$

where α may be positive or negative, a whole number or a fraction.

Expansion of the binomial coefficient is:

$$\binom{\alpha}{n} = \frac{\alpha(\alpha - 1)(\alpha - 2)(\alpha - 3) \ldots (\alpha - n + 1)}{1 \times 2 \times 3 \times \ldots \times n}$$

TAYLOR SERIES

$$f(x) = f(\alpha) + \frac{f'(\alpha)}{1!} (x - \alpha) + \frac{f''(\alpha)}{2!} (x - \alpha)^2 + \ldots$$

For $\alpha = 0$, obtain the Maclaurin Series —

$$f(x) = f(0) + \frac{f'(0)}{1!} x + \frac{f''(0)}{2!} x^2 + \ldots$$

Table 1 of the Appendix gives various formulas for the Taylor series.

Consider the nth degree polynomial

$$P(x) = b_0 + b_1 x + \ldots + b_n x^n$$

If we are interested in $P(x)$ in the neighborhood of some value $x = a$,

$$P(a + h) = b_0 + b_1 (a + h) + \ldots + b_n (a + h)^n$$

which we could expand and write as a polynomial in h:

$$P(a + h) = c_0 + c_1h + c_2h^2 + \ldots + c_nh^n$$

Since the equation is true for $h = 0$, $C_0 = P(a)$. If the equation is differentiated up to n times and each derivative evaluated for $h = 0$, then

$$
\begin{aligned}
C_0 &= P(a) \\
C_1 &= P'(a) \\
C_2 &= P''(a)/2! \\
&\;\;\vdots \\
C_n &= P^n(a)/n!
\end{aligned}
$$

The equation can thus be written as

$$P(a + h) = P(a) + hP'(a) + \frac{h^2}{2!} P''(a) + \ldots + \frac{h^nP^{(n)}(a)}{n!}$$

Now consider any function $f(x)$ which is differentiable throughout $[a,b]$, $n + 1$ times. $f(x)$ is not necessarily a polynomial of nth degree. Substituting $f(a), f'(a) \ldots, f^{(n)}(a)$ for $P(a), P'(a) \ldots, P^{(n)}(a)$ respectively in the above relation, then $P(a + h)$ could be an approximation for $f(a + h)$, for then $P(x)$ and $f(x)$ have the same values at $x = a$, and their first n derivatives are identical at $x = a$. However, $P(x)$ is not necessarily equal to $f(x)$ at any other point $x \neq a$. We must add another term to account for the difference at the point $x = a + h$ and is zero at $x = a$, $f(a + h) = P(a + h) + R_n$. The term R_n is written as $R_n = h^pR/(pn!)$, where p is an arbitrary positive integer which is chosen later and R is a number that is to be determined such that $f(a + h) = P(a + h) + R_n$.

Then

$$f(a + h) = f(a) + hf'(a) + \ldots + \frac{h^n}{n!} f^{(n)}(a) + \frac{h^pR}{pn!}$$

From Rolle's theroem, $x = a + \theta h$, $0 \leq \theta \leq 1$ and R is

$$R = h^{n-p+1} (1 - \theta)^{n-p+1}f^{(n+1)}(a + \theta h)$$

The remainder term is

$$R_n = \frac{h^{n+1}}{pn!} (1 - \theta)^{n-p+1}f^{(n+1)}(a + \theta h)$$

For $p = n + 1$, Lagrange's formula for the remainder results

$$R_n = \frac{h^{n+1}}{(n + 1)!} f^{(n+1)}(a + \theta h)$$

For $p = 1$, we have Cauchy's formula

$$R_n = \frac{h^{n+1}}{n!} (1 - \theta)^n f^{(n+1)}(a + \theta h)$$

The above gives Taylor series with the remainder.

SIMULTANEOUS EQUATIONS

IN TWO UNKNOWNS

Let

$$ax + by = e$$

and

$$cx + dy = f$$

be a system of two equations in two unknowns. Cramer's rule is used to find the solution.

$$x = \frac{\begin{vmatrix} e & b \\ f & d \end{vmatrix}}{\begin{vmatrix} a & b \\ c & d \end{vmatrix}} = \frac{ed - bf}{ad - bc}$$

$$y = \frac{\begin{vmatrix} a & e \\ c & f \end{vmatrix}}{\begin{vmatrix} a & b \\ c & d \end{vmatrix}} = \frac{af - ec}{ad - bc}$$

IN THREE UNKNOWNS

Let

$$a_1x + b_1y + c_1z = d_1$$

$$a_2x + b_2y + c_2z = d_2$$

$$a_3x + b_3y + c_3z = d_3$$

be a system of three equations in three unknowns. Cramer's rule is used to find the solution.

$$x = \frac{\begin{vmatrix} d_1 & b_1 & c_1 \\ d_2 & b_2 & c_2 \\ d_3 & b_3 & c_3 \end{vmatrix}}{\text{Det } A} \qquad y = \frac{\begin{vmatrix} a_1 & d_1 & c_1 \\ a_2 & d_2 & c_2 \\ a_3 & d_3 & c_3 \end{vmatrix}}{\text{Det } A} \qquad z = \frac{\begin{vmatrix} a_1 & b_1 & d_1 \\ a_2 & b_2 & d_2 \\ a_3 & b_3 & d_3 \end{vmatrix}}{\text{Det } A}$$

where $| \quad |$ and Det represent the determinant and

$$A = \begin{bmatrix} a_1 & b_1 & c_1 \\ a_2 & b_2 & c_2 \\ a_3 & b_3 & c_3 \end{bmatrix}$$

SINKING FUND PAYMENT, FUTURE VALUE, NUMBER OF TIME PERIODS

These formulas calculate payment, future value, or number of time periods, given two of the three and the interest rate. Let

n = number of payments
i = periodic interest rate expressed as a decimal, e.g., 6% is expressed as .06
PMT = payment
FV = future value

Then, *FV*, *PMT*, or *n* can be calculated from the other three by the following formulas:

1. $FV = PMT \left[\dfrac{(1 + i)^n - 1}{i} \right]$

2. $PMT = FV \left[\dfrac{i}{(1 + i)^n - 1} \right]$

3. $n = \dfrac{\ln \left(i \dfrac{FV}{PMT} + 1 \right)}{\ln (1 + i)}$

TRIANGLE SOLUTIONS

GIVEN THREE SIDES OF A TRIANGLE (*a*, *b*, *c*)

One can solve the triangle (see Figure A21) for the remaining parameters by the following formulas:

$$C = \cos^{-1} \left(\frac{a^2 + b^2 - c^2}{2ab} \right)$$

$$B = \sin^{-1} \left(\frac{b \sin C}{c} \right)$$

$$A = \sin^{-1} \left(\frac{a \sin C}{c} \right)$$

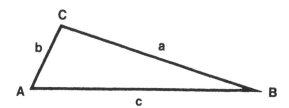

Figure A21. Triangle solution (*a*, *b*, *c*).

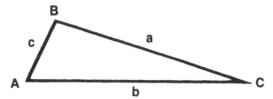

Figure A22. Triangle solution (a, B, C).

Example:

Given the following three sides:

$$a = 30.3$$
$$b = 40.4$$
$$c = 62.6$$

solve the triangle.

Solution:

$$C = 123.99°$$
$$B = 32.35°$$
$$A = 23.66°$$

GIVEN TWO ANGLES AND THEIR INCLUDED SIDE (a, B, C)

One can solve the triangle (see Figure A22) for the remaining parameters by the following formulas:

$$A = 2 \sin^{-1} 1 - (B + C) = \pi \text{ radians} - (B + C) = 180° - (B + C)$$

$$= 200 \text{ grads} - (B + C)$$

$$b = \frac{a \sin B}{\sin A}$$

$$c = \frac{a \sin C}{\sin A}$$

Example:

Given the following two angles and their included side:

$$a = 25.2$$
$$B = 35°20' \text{ (convert } B \text{ and } C \text{ to decimal degrees first)}$$
$$C = 68°30'$$

solve the triangle.

Solution:

$$A = 76.17°$$
$$b = 15.01$$
$$c = 24.15$$

GIVEN TWO SIDES AND THEIR INCLUDED ANGLE (*a, b, C*)

One can solve the triangle (see Figure A23) for the remaining parameters by the following formulas:

$$c = \sqrt{a^2 + c^2 - 2ab \cos C}$$

$$A = \sin^{-1} \left(\frac{a \sin C}{c} \right)$$

$$B = 2 \sin^{-1} 1 - (A + C) = \pi \text{ radians} - (A + C) = 180° - (A + C)$$

$$= 200 \text{ grads} - (A + C)$$

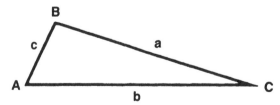

Figure A23. Triangle solution (*a, b, C*).

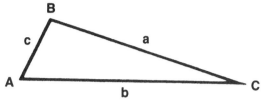

Figure A24. Triangle solution (*a*, *A*, *C*).

Example:

Given the following two sides and included angle:

$$a = 132$$
$$b = 224$$
$$C = 28°40' \text{ (convert to decimal degrees first)}$$

solve the triangle.

Solution:

$$c = 125.35$$
$$A = 30.34°$$
$$B = 120.99°$$

GIVEN TWO ANGLES AND AN OPPOSITE SIDE (*a*, *A*, *C*)

One can solve the triangle (see Figure A24) for the remaining parameters by the following formulas:

$$B = 2 \sin^{-1} 1 - (A + C) = \pi \text{ radians} - (A + C) = 180° - (A + C)$$

$$= 200 \text{ grads} - (A + C)$$

$$b = \frac{a \sin B}{\sin A}$$

$$c = \frac{a \sin C}{\sin A}$$

Example:

Given the following two angles and opposite side:

$$a = 17.5$$
$$A = 41.23°$$
$$C = 62.20°$$

solve the triangle.

Solution:

$$B = 76.57°$$
$$b = 25.83$$
$$c = 23.49$$

GIVEN TWO SIDES AND A NON-INCLUDED ANGLE (*b*, *c*, *B*)

One can solve the triangle (see Figure A25) for the remaining parameters by the following formulas:

1. $C = \sin^{-1}\left(\dfrac{c \sin B}{b}\right)$

2. $A = 2 \sin^{-1} 1 - (B + C)$

    $= \pi \text{ radians} - (B + C) = 180° - (B + C)$

    $= 200 \text{ grads} - (B + C)$

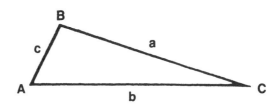

Figure A25. Triangle solution (*B*, *b*, *c*).

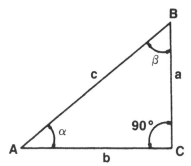

Figure A26. Right triangle *ABC.*

3. $a = \dfrac{b \sin A}{\sin B}$

If *B* is acute (<90°) and *b* < *c*, a second set of solutions exists and is calculated by the following formulas:

4. $C' = 2 \sin^{-1} 1 - C$
5. $A' = 2 \sin^{-1} 1 - (B + C')$

TRIGONOMETRIC RELATIONSHIPS

PLANE TRIGONOMETRY

TRIGONOMETRIC FUNCTIONS OF ACUTE ANGLES

Refer to triangle *ABC* in Figure A26 for which the following definitions apply:

$$\text{sine } \alpha = \sin \alpha = \frac{\text{side opposite}}{\text{hypotenuse}} = \frac{a}{c}$$

$$\text{cosine } \alpha = \cos \alpha = \frac{\text{side adjacent}}{\text{hypotenuse}} = \frac{b}{c}$$

$$\text{tangent } \alpha = \tan \alpha = \frac{\text{side opposite}}{\text{side adjacent}} = \frac{a}{b}$$

$$\text{cotangent } \alpha = \text{ctn } \alpha = \cot \alpha = \frac{\text{side adjacent}}{\text{side opposite}} = \frac{b}{a}$$

$$\text{secant } \alpha = \sec \alpha = \frac{\text{hypotenuse}}{\text{side adjacent}} = \frac{c}{b}$$

$$\text{cosecant } \alpha = \csc \alpha = \frac{\text{hypotenuse}}{\text{side opposite}} = \frac{c}{a}$$

Complementary Relations for Angles α and β

$$\sin \alpha = \cos \beta$$
$$\cos \alpha = \sin \beta$$
$$\tan \alpha = \cot \beta$$
$$\alpha + \beta = 90°$$
$$\cot \alpha = \tan \beta$$
$$\sec \alpha = \csc \beta$$
$$\csc \alpha = \sec \beta$$

TRIGONOMETRIC IDENTITIES

Formulas for Product Relations

$$\sin \alpha = \tan \alpha \cos \alpha$$
$$\cos \alpha = \cot \alpha \sin \alpha$$
$$\tan \alpha = \sin \alpha \sec \alpha$$
$$\sec \alpha = \csc \alpha \tan \alpha$$
$$\csc \alpha = \sec \alpha \cot \alpha$$
$$\cot \alpha = \cos \alpha \csc \alpha$$

Formulas for Function Sum and Difference

$$\sin \alpha + \sin \beta = 2 \sin \tfrac{1}{2}(\alpha + \beta) \cos \tfrac{1}{2}(\alpha - \beta)$$

$$\sin \alpha - \sin \beta = 2 \cos \tfrac{1}{2}(\alpha + \beta) \sin \tfrac{1}{2}(\alpha - \beta)$$

$$\cos \alpha + \cos \beta = 2 \cos \tfrac{1}{2}(\alpha + \beta) \cos \tfrac{1}{2}(\alpha - \beta)$$

$$\cos \alpha - \cos \beta = 2 \sin \tfrac{1}{2}(\alpha + \beta) \sin \tfrac{1}{2}(\alpha - \beta)$$

$$\tan \alpha + \tan \beta = \frac{\sin (\alpha + \beta)}{\cos \alpha \cos \beta}$$

$$\tan \alpha - \tan \beta = \frac{\sin (\alpha - \beta)}{\cos \alpha \cos \beta}$$

$$\cot \alpha + \cot \beta = \frac{\sin (\alpha + \beta)}{\sin \alpha \sin \beta}$$

$$\cot \alpha - \cot \beta = \frac{\sin (\beta - \alpha)}{\sin \alpha \sin \beta}$$

$$\frac{\sin \alpha + \sin \beta}{\sin \alpha - \sin \beta} = \frac{\tan \frac{1}{2}(\alpha + \beta)}{\tan \frac{1}{2}(\alpha - \beta)}$$

$$\frac{\sin \alpha + \sin \beta}{\cos \alpha - \cos \beta} = \cot \frac{1}{2}(\beta - \alpha)$$

$$\frac{\sin \alpha - \sin \beta}{\cos \alpha + \cos \beta} = \tan \frac{1}{2}(\alpha - \beta)$$

FORMULAS FOR QUOTIENT RELATIONS

$$\sin \alpha = \frac{\tan \alpha}{\sec \alpha} \qquad \cos \alpha = \frac{\cot \alpha}{\csc \alpha}$$

$$\tan \alpha = \frac{\sin \alpha}{\cos \alpha} \qquad \csc \alpha = \frac{\sec \alpha}{\tan \alpha}$$

$$\sec \alpha = \frac{\csc \alpha}{\cot \alpha} \qquad \cot \alpha = \frac{\cos \alpha}{\sin \alpha}$$

PYTHAGOREAN FORMULAS

$$\sin^2 \alpha + \cos^2 \alpha = 1$$

$$1 + \tan^2 \alpha = \sec^2 \alpha$$

$$1 + \cot^2 \alpha = \csc^2 \alpha$$

RELATIONS FOR THE SUM AND DIFFERENCE BETWEEN TWO ANGLES

$$\sin (\alpha + \beta) = \sin \alpha \cos \beta + \cos \alpha \sin \beta$$

$$\sin (\alpha - \beta) = \sin \alpha \cos \beta - \cos \alpha \sin \beta$$

$$\cos (\alpha + \beta) = \cos \alpha \cos \beta - \sin \alpha \sin \beta$$

$$\cos (\alpha - \beta) = \cos \alpha \cos \beta + \sin \alpha \sin \beta$$

$$\tan (\alpha + \beta) = \frac{\tan \alpha + \tan \beta}{1 - \tan \alpha \tan \beta}$$

$$\tan (\alpha - \beta) = \frac{\tan \alpha - \tan \beta}{1 + \tan \alpha \tan \beta}$$

$$\cot (\alpha + \beta) = \frac{\cot \beta \cot \alpha - 1}{\cot \beta + \cot \alpha}$$

$$\cot (\alpha - \beta) = \frac{\cot \beta \cot \alpha + 1}{\cot \beta - \cot \alpha}$$

$$\sin (\alpha + \beta) \sin (\alpha - \beta) = \sin^2 \alpha - \sin^2 \beta = \cos^2 \beta - \cos^2 \alpha$$

$$\cos (\alpha + \beta) \cos (\alpha - \beta) = \cos^2 \alpha - \sin^2 \beta = \cos^2 \beta - \sin^2 \alpha$$

FORMULAS FOR DOUBLE ANGLES

$$\sin 2\alpha = 2 \sin \alpha \cos \alpha = \frac{2 \tan \alpha}{1 + \tan^2 \alpha}$$

$$\cos 2\alpha = \cos^2 - \alpha \sin^2 \alpha = 2 \cos^2 \alpha - 1 = 1 - 2 \sin^2 \alpha = \frac{1 - \tan^2 \alpha}{1 + \tan^2 \alpha}$$

$$\tan 2\alpha = \frac{2 \tan \alpha}{1 - \tan^2 \alpha}$$

$$\cot 2\alpha = \frac{\cot^2 \alpha - 1}{2 \cot \alpha}$$

Formulas for Product Functions

$$\sin \alpha \sin \beta = \tfrac{1}{2}\cos (\alpha - \beta) - \tfrac{1}{2}\cos (\alpha + \beta)$$

$$\cos \alpha \cos \beta = \tfrac{1}{2}\cos (\alpha - \beta) + \tfrac{1}{2}\cos (\alpha + \beta)$$

$$\sin \alpha \cos \beta = \tfrac{1}{2}\sin (\alpha + \beta) + \tfrac{1}{2}\sin (\alpha - \beta)$$

$$\cos \alpha \sin \beta = \tfrac{1}{2}\sin (\alpha + \beta) - \tfrac{1}{2}\sin (\alpha - \beta)$$

Formulas for Power Relations

$$\sin^2 \alpha = \tfrac{1}{2}(1 - \cos 2\alpha)$$

$$\sin^3 \alpha = \tfrac{1}{4}(3 \sin \alpha - \sin 3\alpha)$$

$$\sin^4 \alpha = \tfrac{1}{8}(3 - 4 \cos 2\alpha + \cos 4\alpha)$$

$$\cos^2 \alpha = \tfrac{1}{2}(1 + \cos 2\alpha)$$

$$\cos^3 \alpha = \tfrac{1}{4}(3 \cos \alpha + \cos 3\alpha)$$

$$\cos^4 \alpha = \tfrac{1}{8}(3 + 4 \cos 2\alpha + \cos 4\alpha)$$

$$\tan^2 \alpha = \frac{1 - \cos 2\alpha}{1 + \cos 2\alpha}$$

$$\cot^2 \alpha = \frac{1 + \cos 2\alpha}{1 - \cos 2\alpha}$$

Formulas for Multiple Angles

$$\sin 3\alpha = 3 \sin \alpha - 4 \sin^3 \alpha$$

$$\cos 3\alpha = 4 \cos^3 \alpha - 3 \cos \alpha$$

$$\sin 4\alpha = 4 \sin \alpha \cos \alpha - 8 \sin^3 \alpha \cos \alpha$$

$$\cos 4\alpha = 8 \cos^4 \alpha - 8 \cos^2 \alpha + 1$$

$$\sin 5\alpha = 5 \sin \alpha - 20 \sin^3 \alpha + 16 \sin^5 \alpha$$

$$\cos 5\alpha = 16 \cos^5 \alpha - 20 \cos^3 \alpha + 5 \cos \alpha$$

$$\sin \eta\alpha = 2 \sin (\eta - 1)\alpha \cos \alpha - \sin (\eta - 2)\alpha$$

$$\cos \eta\alpha = 2 \cos (\eta - 1)\alpha \cos \alpha - \cos (\eta - 2)\alpha$$

$$\tan 3\alpha = \frac{3 \tan \alpha - \tan^3 \alpha}{1 - 3 \tan^2 \alpha}$$

$$\tan 4\alpha = \frac{4 \tan \alpha - 4 \tan^3 \alpha}{1 - 6 \tan^2 \alpha + \tan^4 \alpha}$$

$$\tan \eta\alpha = \frac{\tan [(\eta - 1) \alpha] + \tan \alpha}{1 - \tan [(\eta - 1) \alpha] \tan \alpha}$$

Formulas for Half Angles

$$\sin \frac{\alpha}{2} = \pm \sqrt{\frac{1 - \cos \alpha}{2}}$$

$$\cos \frac{\alpha}{2} = \pm \sqrt{\frac{1 + \cos \alpha}{2}}$$

$$\tan \frac{\alpha}{2} = \pm \sqrt{\frac{1 - \cos \alpha}{1 + \cos \alpha}} = \frac{1 - \cos \alpha}{\sin \alpha} = \frac{\sin \alpha}{1 + \cos \alpha}$$

$$\cot \frac{\alpha}{2} = \pm \sqrt{\frac{1 + \cos \alpha}{1 - \cos \alpha}} = \frac{1 + \cos \alpha}{\sin \alpha} = \frac{\sin \alpha}{1 - \cos \alpha}$$

RELATIONSHIPS AMONG TRIGONOMETRIC FUNCTIONS

Relationships among trigonometric functions.

Function	sin α	cos α	tan α
sin α	sin α	$\pm \sqrt{1 - \cos^2 \alpha}$	$\dfrac{\tan \alpha}{\pm \sqrt{1 + \tan^2 \alpha}}$

Relationships among trigonometric functions *(continued)*.

Function	$\sin \alpha$	$\cos \alpha$	$\tan \alpha$
$\cos \alpha$	$\pm \sqrt{1 - \sin^2 \alpha}$	$\cos \alpha$	$\dfrac{1}{\pm \sqrt{1 + \tan^2 \alpha}}$
$\tan \alpha$	$\dfrac{\sin \alpha}{\pm \sqrt{1 - \sin^2 \alpha}}$	$\dfrac{\pm \sqrt{1 - \cos^2 \alpha}}{\cos \alpha}$	$\tan \alpha$
$\cot \alpha$	$\dfrac{\pm \sqrt{1 - \sin^2 \alpha}}{\sin \alpha}$	$\dfrac{\cos \alpha}{\pm \sqrt{1 - \cos^2 \alpha}}$	$\dfrac{1}{\tan \alpha}$
$\sec \alpha$	$\dfrac{1}{\pm \sqrt{1 - \sin^2 \alpha}}$	$\dfrac{1}{\cos \alpha}$	$\pm \sqrt{1 + \tan^2 \alpha}$
$\csc \alpha$	$\dfrac{1}{\sin \alpha}$	$\dfrac{1}{\pm \sqrt{1 - \cos^2 \alpha}}$	$\dfrac{\pm \sqrt{1 + \tan^2 \alpha}}{\tan \alpha}$

Relationships among trigonometric functions.

Function	$\cot \alpha$	$\sec \alpha$	$\csc \alpha$
$\sin \alpha$	$\dfrac{1}{\pm \sqrt{1 + \cot^2 \alpha}}$	$\dfrac{\pm \sqrt{\sec^2 \alpha - 1}}{\sec \alpha}$	$\dfrac{1}{\csc \alpha}$
$\cos \alpha$	$\dfrac{\cot \alpha}{\pm \sqrt{1 + \cot^2 \alpha}}$	$\dfrac{1}{\sec \alpha}$	$\dfrac{\pm \sqrt{\csc^2 \alpha - 1}}{\csc \alpha}$
$\tan \alpha$	$\dfrac{1}{\cot \alpha}$	$\pm \sqrt{\sec^2 \alpha - 1}$	$\dfrac{1}{\pm \sqrt{\csc^2 \alpha - 1}}$
$\cot \alpha$	$\cot \alpha$	$\dfrac{1}{\pm \sqrt{\sec^2 \alpha - 1}}$	$\pm \sqrt{\csc^2 \alpha}$

(continued)

Relationships among trigonometric functions *(continued)*.

Function	cot α	sec α	csc α
sec α	$\dfrac{\pm \sqrt{1 + \cot^2 \alpha}}{\cot \alpha}$	sec α	$\dfrac{\csc \alpha}{\pm \sqrt{\csc^2 \alpha - 1}}$
csc α	$\pm \sqrt{1 + \cot^2 \alpha}$	$\dfrac{\sec \alpha}{\pm \sqrt{\sec^2 \alpha - 1}}$	csc α

VERSINE, COVERSINE, HAVERSINE, EXSECANT

The above are calculated by the following formulas:

1. versine (versed sine)

$$\text{vers } \theta = 1 - \cos \theta$$

2. coversine (covered sine)

$$\text{cov } \theta = 1 - \sin \theta$$

3. haversine

$$\text{hav } \theta = \tfrac{1}{2} \text{ vers } \theta = \sin^2 \tfrac{1}{2}\theta$$

4. exsecant

$$\text{exsec } \theta = \sec \theta - 1$$

SPHERICAL TRIGONOMETRY

FORMULAS AND DEFINITIONS FOR OBLIQUE SPHERICAL TRIANGLES (Figure A27)

Law of Sines

$$\frac{\sin a}{\sin A} = \frac{\sin b}{\sin B} = \frac{\sin c}{\sin C}$$

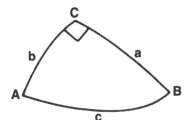

Figure A27. Spherical triangle.

where a, b, c represent the sides of any spherical triangle and A, B, C represent the corresponding opposite angles

Law of Cosines for Sides

$$\cos a = \cos b \cos c + \sin b \sin c \cos A$$

$$\cos b = \cos c \cos a + \sin c \sin a \cos B$$

$$\cos c = \cos a \cos b + \sin a \sin b \cos C$$

Law of Tangents

$$\frac{\tan \tfrac{1}{2}(B - C)}{\tan \tfrac{1}{2}(B + C)} = \frac{\tan \tfrac{1}{2}(b - c)}{\tan \tfrac{1}{2}(b + c)}$$

$$\frac{\tan \tfrac{1}{2}(C - A)}{\tan \tfrac{1}{2}(C + A)} = \frac{\tan \tfrac{1}{2}(c - a)}{\tan \tfrac{1}{2}(c + a)}$$

$$\frac{\tan \tfrac{1}{2}(A - B)}{\tan \tfrac{1}{2}(A + B)} = \frac{\tan \tfrac{1}{2}(a - b)}{\tan \tfrac{1}{2}(a + b)}$$

Gauss's Formulae

$$\frac{\sin \tfrac{1}{2}(a - b)}{\sin \tfrac{1}{2}c} = \frac{\sin \tfrac{1}{2}(A - B)}{\sin \tfrac{1}{2}C}$$

$$\frac{\cos \tfrac{1}{2}(a - b)}{\cos \tfrac{1}{2}c} = \frac{\sin \tfrac{1}{2}(A + B)}{\cos \tfrac{1}{2}C}$$

$$\frac{\sin \frac{1}{2}(a + b)}{\sin \frac{1}{2}c} = \frac{\cos \frac{1}{2}(A - B)}{\sin \frac{1}{2}C}$$

$$\frac{\cos \frac{1}{2}(a + b)}{\cos \frac{1}{2}c} = \frac{\cos \frac{1}{2}(A + B)}{\sin \frac{1}{2}C}$$

VECTOR PRODUCTS

Let

$$\bar{a} = a_1\bar{i} + a_2\bar{j} + a_3\bar{k} \quad \text{and} \quad b = b_1\bar{i} + b_2\bar{j} + b_3\bar{k}$$

where $\bar{i}, \bar{j}, \bar{k}$ are unit vectors parallel to the x-, y-, z-axes, respectively. The magnitude of $\bar{a}$ is $|\bar{a}| = + (a_1^2 + a_2^2 + a_3^2)^{1/2}$, similarly $|\bar{b}| = + (b_1^2 + b_2^2 + b_3^2)^{1/2}$.

SCALAR OR DOT PRODUCT

$$\bar{i}\cdot\bar{i} = \bar{j}\cdot\bar{j} = \bar{k}\cdot\bar{k} = 0 \quad \text{and} \quad \bar{i}\cdot\bar{j} = \bar{j}\cdot\bar{k} = \bar{k}\cdot\bar{i} = 1$$

therefore:

$$\bar{a}\cdot\bar{b} = a_1b_1 + a_2b_2 + a_3b_3$$

is the dot product of $\bar{a}$ and $\bar{b}$

also:

$$\bar{a}\cdot\bar{b} = |a|\ |b|\ \cos \theta$$

where θ is the angle between $\bar{a}$ and $\bar{b}$

therefore

$$\theta = \cos^{-1} \frac{(a\cdot b)}{(|\bar{a}|\ |\bar{b}|)} = \frac{a_1b_1 + a_2b_2 + a_3b_3}{(a_1^2 + a_2^2 + a_3^2)^{1/2}\ (b_1^2 + b_2^2 + b_3^2)^{1/2}}$$

VECTOR OR CROSS PRODUCT

$$\bar{i} \times \bar{i} = \bar{j} \times \bar{j} = \bar{k} \times \bar{k} = 0$$

$$\bar{i} \times \bar{j} = \bar{k}$$

$$\bar{j} \times \bar{k} = \bar{i}$$

$$\bar{k} \times \bar{i} = \bar{j}$$

$$\bar{j} \times \bar{i} = -\bar{k}$$

$$\bar{k} \times \bar{j} = -\bar{i}$$

$$\bar{i} \times \bar{k} = -\bar{j}$$

Vector multiplication is simplified by the use of a 3×3 determinant:

$$\bar{c} = \bar{a} \times \bar{b} = \begin{vmatrix} \bar{i} & \bar{j} & \bar{k} \\ a_1 & a_2 & a_3 \\ b_1 & b_2 & b_3 \end{vmatrix}$$

$$= (a_2 b_3 - a_3 b_2)\bar{i} - (a_1 b_3 - a_3 b_1)\bar{j} + (a_1 b_2 - a_2 b_1)\bar{k}$$

is the cross product of $\bar{a}$ and $\bar{b}$. $\bar{c}$ is also a vector perpendicular to the plane of $\bar{a}$ and $\bar{b}$ in the sense of a right-handed screw.

The absolute value of $\bar{c} = |\bar{a} \times \bar{b}| = |\bar{a}| \, |\bar{b}| \sin \theta$

Example:

Find the cross product of the two vectors $\overline{A} = 2.34 \, \bar{i} + 5.17 \, \bar{j} + 7.43 \, \bar{k}$ and $\overline{B} = .072 \, \bar{i} + .231 \, \bar{j} + .409 \, \bar{k}$.

Solution:

$$\overline{C} = \overline{A} \times \overline{B} = .40 \, \bar{i} - .42 \, \bar{j} + .17 \, \bar{k}$$

APPENDIX A—TABLES OF FUNCTIONS

Table A1. Formulas for derivatives.

$$\frac{d}{dx}(a) = 0$$

$$\frac{d}{dx}(x) = 1$$

$$\frac{d}{dx}(au) = a\,\frac{du}{dx}$$

$$\frac{d}{dx}(u + v - z) = \frac{dv}{dx} + \frac{dv}{dx} - \frac{dz}{dx}$$

$$\frac{d}{dx}(uv) = u\,\frac{dv}{dx} + v\,\frac{du}{dx}$$

$$\frac{d}{dx}(uvz) = uv\,\frac{dz}{dx} + vz\,\frac{du}{dx} + uz\,\frac{dv}{dx}$$

$$\frac{d}{dx}(u^n) = nu^{n-1}\,\frac{du}{dx}$$

$$\frac{d}{dx}\left(\frac{1}{u}\right) = -\frac{1}{u^2}\,\frac{du}{dx}$$

$$\frac{d}{dx}(u^{-n}) = -\frac{n}{u^{n+1}}\,\frac{du}{dx}$$

$$\frac{d}{dx}(\ln u) = \frac{1}{u}\,\frac{du}{dx}$$

$$\frac{d}{dx}(e^u) = e^u\,\frac{du}{dx}$$

$$\frac{d}{dx}(\sin u) = \frac{du}{dx}(\cos u)$$

$$\frac{d}{dx}(\cos u) = -\frac{du}{dx}(\sin u)$$

$$\frac{d}{dx}(\tan u) = \frac{du}{dx}(\sec^2 u)$$

$$\frac{d}{dx}(\cot u) = -\frac{du}{dx}(\csc^2 u)$$

$$\frac{d}{dx}(\sec u) = \frac{du}{dx} \sec u \cdot \tan u$$

$$\frac{d}{dx}(\csc u) = -\frac{du}{dx} \csc u \cdot \cot u$$

$$\frac{d}{dx}(\sinh u) = \frac{du}{dx}(\cosh u)$$

$$\frac{d}{dx}(\cosh u) = \frac{du}{dx}(\sinh u)$$

$$\frac{d}{dx}(\tanh u) = \frac{du}{dx}(\operatorname{sech}^2 u)$$

$$\frac{d}{dx}(\coth u) = -\frac{du}{dx}(\operatorname{csch}^2 u)$$

$$\frac{d}{dx}(\operatorname{sech} u) = -\frac{du}{dx}(\operatorname{sech} u \cdot \tanh u)$$

$$\frac{d}{dx}(\operatorname{csch} u) = -\frac{du}{dx}(\operatorname{csch} u \cdot \coth u)$$

$$\frac{d}{dx}(\sinh^{-1} u) = \frac{1}{\sqrt{u^2 + 1}} \frac{du}{dx}$$

$$\frac{d}{dx}(\tanh^{-1} u) = \frac{1}{1 - u^2} \frac{du}{dx} \ , \ (u^2 < 1)$$

$$\frac{d}{dx}(\coth^{-1} u) = \frac{1}{1 - u^2} \frac{du}{dx} \ , \ (u^2 > 1)$$

$$\frac{d}{dx}(\operatorname{csch}^{-1} u) = -\frac{1}{|u| \sqrt{u^2 + 1}} \frac{du}{dx}$$

$$\frac{d}{dq} \int_p^q f(x)dx = f(q), \ [p = \text{constant}]$$

$$\frac{d}{dp} \int_p^q f(x)dx = -f(p), \ [q = \text{constant}]$$

Note that u, v, z represent functions of x and a, n represent fixed real numbers.

Table A2. General formulas for integrals.

Equations 1–13 give elementary forms; Equations 14–25 give forms containing $(a + bx)$; Equations 26–32 give forms containing $\sqrt{a + bx}$; Equations 33–40 give forms containing $\sqrt{x^2 \pm a^2}$.

1. $\displaystyle\int a\,dx = ax$

2. $\displaystyle\int a \cdot f(x)\,dx = a \int f(x)\,dx$

3. $\displaystyle\int (u + v)\,dx = \int u\,dx + \int v\,dx$ (note u and v are functions of x)

4. $\displaystyle\int u\,dv = u \int dv - \int v\,du = uv - \int v\,du$

5. $\displaystyle\int \frac{f'(x)\,dx}{f(x)} = \ln f(x)$ [where $df(x) = f'(x)\,dx$]

6. $\displaystyle\int \frac{dx}{x} = \ln x$

7. $\displaystyle\int e^x\,dx = e^x$

8. $\displaystyle\int e^{ax}\,dx = e^{ax}/a$

9. $\displaystyle\int \ln x\,dx = x \ln x - x$

10. $\displaystyle\int a^x \ln a\,dx = a^x$ (for $a > 0$)

11. $\displaystyle\int \frac{dx}{a^2 + x^2} = \frac{1}{a} \tan^{-1} \frac{x}{a}$

12. $\displaystyle\int \frac{dx}{\sqrt{x^2 \pm a^2}} = \ln (x + \sqrt{x^2 \pm a^2})$

13. $\displaystyle\int \frac{dx}{x\sqrt{a^2 \pm x^2}} = -\frac{1}{a} \ln \left(\frac{a + \sqrt{a^2 \pm x^2}}{x} \right)$

14. $\displaystyle\int (a + bx)^n\,dx = \frac{(a + bx)^{n+1}}{(n + 1)b}$ $(n \neq -1)$

68

15. $\displaystyle \int x^2 (a + bx)^n \, dx = \frac{1}{b^3} \left[\frac{(a + bx)^{n+3}}{n + 3} - 2a \, \frac{(a + bx)^{n+2}}{n + 2} + a^2 \, \frac{(a + bx)^{n+1}}{n + 1} \right]$

16. $\displaystyle \int \frac{dx}{a + bx} = \frac{1}{b} \, \ln (a + bx)$

17. $\displaystyle \int \frac{dx}{(a + bx)^2} = - \frac{1}{b(a + bx)}$

18. $\displaystyle \int \frac{dx}{(a + bx)^3} = - \frac{1}{2b(a + bx)^2}$

19. $\displaystyle \int \frac{x \, dx}{(a + bx)^2} = \frac{1}{b^2} \left[\ln (a + bx) + \frac{a}{a + bx} \right]$

20. $\displaystyle \int \frac{x^2 \, dx}{(a + bx)^2} = \frac{1}{b^2} \left[a + bx - 2a \ln (a + bx) - \frac{a^2}{a + bx} \right]$

21. $\displaystyle \int \frac{x^2 \, dx}{(a + bx)^3} = \frac{1}{b^3} \left[\ln (a + bx) + \frac{2a}{a + bx} - \frac{a^2}{2(a + bx)^2} \right]$

22. $\displaystyle \int \frac{dx}{x(a + bx)} = - \frac{1}{a} \, \ln \frac{a + bx}{x}$

23. $\displaystyle \int \frac{dx}{x^2 (a + bx)} = - \frac{1}{ax} + \frac{b}{a^2} \, \ln \frac{a + bx}{x}$

24. $\displaystyle \int \frac{dx}{x^3 (a + bx)} = \frac{2bx - a}{2a^2 x^2} + \frac{b^2}{a^3} \, \ln \frac{x}{a + bx}$

25. $\displaystyle \int \frac{dx}{x^2 (a + bx)^2} = - \frac{a + 2bx}{a^2 x (a + bx)} + \frac{2b}{a^3} \, \ln \frac{a + bx}{x}$

26. $\displaystyle \int \sqrt{a + bx} \, dx = \frac{2}{3b} \, \sqrt{(a + bx)^3}$

27. $\displaystyle \int x \sqrt{a + bx} \, dx = - \frac{2 (2a - 3bx) \sqrt{(a + bx)^3}}{15b^2}$

28. $\displaystyle \int \frac{\sqrt{a + bx}}{x} \, dx = 2\sqrt{a + bx} + a \int \frac{dx}{x\sqrt{a + bx}}$

29. $\displaystyle \int \frac{\sqrt{a + bx}}{x^2} \, dx = - \frac{\sqrt{a + bx}}{x} + \frac{b}{2} \int \frac{dx}{x\sqrt{a + bx}}$

(continued)

30. $\displaystyle\int \frac{dx}{\sqrt{a+bx}} = \frac{2\sqrt{a+bx}}{b}$

31. $\displaystyle\int \frac{xdx}{\sqrt{a+bx}} = -\frac{2(2a-bx)}{3b^2}\sqrt{a+bx}$

32. $\displaystyle\int \frac{x^2dx}{\sqrt{a+bx}} = \frac{2(8a^2-4abx+3b^2x^2)}{15b^3}\sqrt{a+bx}$

33. $\displaystyle\int \frac{dx}{\sqrt{x^2\pm a^2}} = \log(x+\sqrt{x^2\pm a^2})$

34. $\displaystyle\int \frac{dx}{x\sqrt{x^2-a^2}} = \frac{1}{|a|}\sec^{-1}\frac{x}{a}$

35. $\displaystyle\int \frac{dx}{x\sqrt{x^2+a^2}} = -\frac{1}{a}\log\left(\frac{a+\sqrt{x^2+a^2}}{x}\right)$

36. $\displaystyle\int \frac{xdx}{\sqrt{x^2\pm a^2}} = \sqrt{x^2\pm a^2}$

37. $\displaystyle\int x\sqrt{x^2\pm a^2}\,dx = \frac{1}{3}\sqrt{(x^2\pm a^2)^3}$

38. $\displaystyle\int \frac{dx}{\sqrt{(x^2\pm a^2)^3}} = \frac{\pm x}{a^2\sqrt{x^2\pm a^2}}$

39. $\displaystyle\int x\sqrt{(x^2\pm a^2)^3}\,dx = \frac{1}{5}\sqrt{(x^2\pm a^2)^5}$

40. $\displaystyle\int \frac{dx}{(x+a)\sqrt{x^2-a^2}} = \frac{\sqrt{x^2-a^2}}{a(x+a)}$

Table A3. Integrals of trigonometric, inverse trigonometric and hyperbolic functions.

Trigonometric Functions

1. $\displaystyle\int (\sin ax)\,dx = -\frac{1}{a}\cos ax$

2. $\displaystyle\int (\cos ax)\,dx = \frac{1}{a}\sin ax$

3. $\displaystyle\int (\tan ax)\,dx = -\frac{1}{a}\ln\cos ax = \frac{1}{a}\ln\sec ax$

70

4. $\displaystyle\int (\cot ax)\ dx = \frac{1}{a}\ \ln \sin ax = -\ \frac{1}{a}\ \ln \csc ax$

5. $\displaystyle\int (\sec ax)\ dx = \frac{1}{a}\ \ln (\sec ax + \tan ax) = \frac{1}{a}\ \ln \tan\ \left(\frac{\pi}{4} + \frac{ax}{2}\right)$

6. $\displaystyle\int (\csc ax)\ dx = \frac{1}{a}\ \ln (\csc ax - \cot ax) = \frac{1}{a}\ \ln \tan\ \frac{ax}{2}$

7. $\displaystyle\int (\sin^2 ax)\ dx = -\ \frac{1}{2a}\ \cos ax \sin ax + \frac{1}{2}\ x = \frac{1}{2}\ x - \frac{1}{4a}\ \sin 2ax$

8. $\displaystyle\int (\sin^3 ax)\ dx = -\ \frac{1}{3a}\ (\cos ax)(\sin^2 ax + 2)$

9. $\displaystyle\int (\sin^n ax)\ dx = -\ \frac{\sin^{n-1} ax \cos ax}{na}\ +\ \frac{n-1}{n}\ \int (\sin^{n-2} ax)\ dx$

10. $\displaystyle\int (\cos^2 ax)\ dx = \frac{1}{2a}\ \sin ax \cos ax + \frac{1}{2}\ x = \frac{1}{2}\ x + \frac{1}{4a}\ \sin 2ax$

11. $\displaystyle\int (\cos^3 ax)\ dx = \frac{1}{3a}\ (\sin ax)(\cos^2 ax + 2)$

12. $\displaystyle\int (\cos^n ax)\ dx = \frac{1}{na}\ \cos^{n-1} ax \sin ax + \frac{n-1}{n}\ \int (\cos^{n-2} ax)\ dx$

13. $\displaystyle\int \frac{dx}{\cos^2 ax} = \int (\sec^2 ax)\ dx = \frac{1}{a}\ \tan ax$

14. $\displaystyle\int \frac{dx}{\cos^n ax} = \int (\sec^n ax)\ dx = \frac{1}{(n-1)a} \cdot \frac{\sin ax}{\cos^{n-1} ax} + \frac{n-2}{n-1}\ \int \frac{dx}{\cos^{n-2} ax}$

15. $\displaystyle\int (\sin ax)(\cos ax)\ dx = \frac{1}{2a}\ \sin^2 ax$

16. $\displaystyle\int (\sin^2 ax)(\cos^2 ax)\ dx = -\ \frac{1}{32a}\ \sin 4ax + \frac{x}{8}$

17. $\displaystyle\int \frac{\sin ax}{\cos^2 ax}\ dx = \frac{1}{a \cos ax} = \frac{1}{a}\ \sec ax$

18. $\displaystyle\int \frac{dx}{(\sin ax)(\cos ax)} = \frac{1}{a}\ \ln \tan ax$

(continued)

19. $\displaystyle \int \frac{dx}{(\sin ax)(\cos^2 ax)} = \frac{1}{a}\left(\sec ax + \ln \tan \frac{ax}{2} \right)$

20. $\displaystyle \int \frac{dx}{(\sin^2 ax)(\cos^2 ax)} = -\frac{2}{a}\cot 2ax$

21. $\displaystyle \int \sin (a + bx)\, dx = -\frac{1}{b}\cos (a + bx)$

22. $\displaystyle \int \cos (a + bx)\, dx = \frac{1}{b}\sin (a + bx)$

23. $\displaystyle \int \frac{dx}{1 + \cos ax} = \frac{1}{a}\tan \frac{ax}{2}$

24. $\displaystyle \int \frac{dx}{1 - \cos ax} = -\frac{1}{a}\cot \frac{ax}{2}$

25. $\displaystyle \int \frac{\sin x\, dx}{a + b \sin x} = \frac{x}{b} - \frac{a}{b}\int \frac{dx}{a + b \sin x}$

26. $\displaystyle \int \frac{dx}{(\sin x)(a + b \sin x)} = \frac{1}{a}\ln \tan \frac{x}{2} - \frac{b}{a}\int \frac{dx}{a + b \sin x}$

27. $\displaystyle \int \frac{\cos ax}{1 + \cos ax}\, dx = x - \frac{1}{a}\tan \frac{ax}{2}$

28. $\displaystyle \int \frac{\cos ax}{1 - \cos ax}\, dx = -x - \frac{1}{a}\cot \frac{ax}{2}$

29. $\displaystyle \int \frac{dx}{(1 + \cos ax)^2} = \frac{1}{2a}\tan \frac{ax}{2} + \frac{1}{6a}\tan^3 \frac{ax}{2}$

30. $\displaystyle \int \frac{dx}{(1 - \cos ax)^2} = \frac{1}{2a}\cot \frac{ax}{2} - \frac{1}{6a}\cot^3 \frac{ax}{2}$

31. $\displaystyle \int x^2 (\sin^2 ax)\, dx = \frac{x^3}{6} - \left(\frac{x}{4a} - \frac{1}{8a^3} \right)\sin 2ax - \frac{x \cos 2ax}{4a^2}$

32. $\displaystyle \int x (\cos^2 ax)\, dx = \frac{x^2}{4} + \frac{x \sin 2ax}{4a} + \frac{\cos 2ax}{8a^2}$

33. $\displaystyle \int x^2 (\cos^2 ax)\, dx = \frac{x^3}{6} + \left(\frac{x^2}{4a} - \frac{1}{8a^3} \right)\sin 2ax + \frac{x \cos 2ax}{4a^2}$

34. $\displaystyle\int x\,(\cos^3 ax)dx = \frac{x\sin 3ax}{12a} + \frac{\cos 3ax}{36a^2} + \frac{3x\sin ax}{4a} + \frac{3\cos ax}{4a^2}$

35. $\displaystyle\int \frac{x}{1 - \cos ax}\,dx = -\frac{x}{a}\cot\frac{ax}{2} + \frac{2}{a^2}\ln\sin\frac{ax}{2}$

36. $\displaystyle\int \frac{x + \sin x}{1 + \cos x}\,dx = x\tan\frac{x}{2}$

37. $\displaystyle\int \frac{x - \sin x}{1 - \cos x}\,dx = -x\cot\frac{x}{2}$

38. $\displaystyle\int (\tan^2 ax)\,dx = \frac{1}{a}\tan ax - x$

39. $\displaystyle\int (\tan^3 ax)\,dx = \frac{1}{2a}\tan^2 ax + \frac{1}{a}\ln\cos ax$

40. $\displaystyle\int (\tan^n ax)\,dx = \frac{\tan^{n-1} ax}{a(n-1)} - \int (\tan^{n-2} ax)\,dx$

Inverse Trigonometric Functions

41. $\displaystyle\int (\sin^{-1} ax)\,dx = x\sin^{-1} ax + \frac{1}{a}\sqrt{1 - a^2x^2}$

42. $\displaystyle\int (\cos^{-1} ax)\,dx = x\cos^{-1} ax - \frac{1}{a}\sqrt{1 - a^2x^2}$

43. $\displaystyle\int (\tan^{-1} ax)\,dx = x\tan^{-1} ax - \frac{1}{2a}\ln(1 + a^2x^2)$

44. $\displaystyle\int (\cot^{-1} ax)\,dx = x\cot^{-1} ax + \frac{1}{2a}\ln(1 + a^2x^2)$

45. $\displaystyle\int (\sec^{-1} ax)\,dx = x\sec^{-1} ax - \frac{1}{a}\ln(ax + \sqrt{a^2x^2 - 1})$

46. $\displaystyle\int (\csc^{-1} ax)\,dx = x\csc^{-1} ax + \frac{1}{a}\ln(ax + \sqrt{a^2x^2 - 1})$

47. $\displaystyle\int \left(\tan^{-1}\frac{x}{a}\right)\,dx = x\tan^{-1}\frac{x}{a} - \frac{a}{2}\ln(a^2 + x^2)$

(continued)

48. $\displaystyle\int \left(\cot^{-1} \frac{x}{a} \right) dx = x \cot^{-1} \frac{x}{a} + \frac{a}{2} \ln (a^2 + x^2)$

49. $\displaystyle\int x (\tan^{-1} ax) dx = \frac{1 + a^2x^2}{2a^2} \tan^{-1} ax - \frac{x}{2a}$

50. $\displaystyle\int x (\cot^{-1} ax) dx = \frac{1 + a^2x^2}{2a^2} \cot^{-1} ax + \frac{x}{2a}$

51. $\displaystyle\int \frac{\cot^{-1} ax}{x^2} dx = -\frac{1}{x} \cot^{-1} ax - \frac{a}{2} \ln \frac{x^2}{a^2x^2 + 1}$

52. $\displaystyle\int x \sec^{-1} ax \, dx = \frac{x^2}{2} \sec^{-1} ax - \frac{1}{2a^2} \sqrt{a^2x^2 - 1}$

53. $\displaystyle\int x \csc^{-1} ax \, dx = \frac{x^2}{2} \csc^{-1} ax + \frac{1}{2a^2} \sqrt{a^2x^2 + 1}$

Hyperbolic Forms

54. $\displaystyle\int (\sinh x) \, dx = \cosh x$

55. $\displaystyle\int (\cosh x) \, dx = \sinh x$

56. $\displaystyle\int (\tanh x) \, dx = \ln \cosh x$

57. $\displaystyle\int (\coth x) \, dx = \ln \sinh x$

58. $\displaystyle\int (\operatorname{sech} x) \, dx = \tan^{-1} (\sinh x)$

59. $\displaystyle\int \operatorname{csch} x \, dx = \ln \tanh \left(\frac{x}{2} \right)$

60. $\displaystyle\int x (\sinh x) \, dx = x \cosh x - \sinh x$

61. $\displaystyle\int x^n (\sinh x) \, dx = x^n \cosh x - n \int x^{n-1} (\cosh x) \, dx$

62. $\displaystyle\int x\,(\cosh x)\,dx = x\,\sinh x - \cosh x$

63. $\displaystyle\int x^n\,(\cosh x)\,dx = x^n\,\sinh x - n\int x^{n-1}\,(\sinh x)\,dx$

64. $\displaystyle\int (\operatorname{sech} x)\,(\tanh x)\,dx = -\operatorname{sech} x$

65. $\displaystyle\int (\operatorname{csch} x)\,(\coth x)\,dx = -\operatorname{csch} x$

66. $\displaystyle\int (\sinh^2 x)\,dx = \frac{\sinh 2x}{4} - \frac{x}{2}$

67. $\displaystyle\int (\tanh^2 x)\,dx = x - \tanh x$

68. $\displaystyle\int (\operatorname{sech}^2 x)\,dx = \tanh x$

69. $\displaystyle\int (\coth^2 x)\,dx = x - \coth x$

70. $\displaystyle\int (\operatorname{csch}^2 x)\,dx = -\coth x$

Table A4. Integrals of logarithmic and exponential forms.

Logarithmic Forms

1. $\displaystyle\int (\ln x)\,dx = x\,\ln x - x$

2. $\displaystyle\int x\,(\ln x)\,dx = \frac{x^2}{2}\,\ln x - \frac{x^2}{4}$

3. $\displaystyle\int x^2\,(\ln x)\,dx = \frac{x^3}{3}\,\ln x - \frac{x^3}{9}$

(continued)

4. $\int x^n (\ln ax) \, dx = \dfrac{x^{n+1}}{n+1} \ln ax - \dfrac{x^{n+1}}{(n+1)^2}$

5. $\int (\ln x)^2 \, dx = x (\ln x)^2 - 2x \ln x + 2x$

6. $\int \dfrac{dx}{x \ln x} = \ln (\ln x)$

7. $\int \sin (\ln x) \, dx = \dfrac{1}{2} x \sin (\ln x) - \dfrac{1}{2} x \cos (\ln x)$

8. $\int \cos (\ln x) \, dx = \dfrac{1}{2} x \sin (\ln x) + \dfrac{1}{2} x \cos (\ln x)$

9. $\int [\ln (x^2 + a^2)] \, dx = x \ln (x^2 + a^2) - 2x + 2a \tan^{-1} \dfrac{x}{a}$

10. $\int [\ln (x^2 - a^2)] \, dx = x \ln (x^2 - a^2) - 2x + 2a \ln \dfrac{x+a}{x-a}$

Exponential Forms

11. $\int e^x \, dx = e^x$

12. $\int e^{-x} \, dx = -e^{-x}$

13. $\int e^{ax} \, dx = \dfrac{e^{ax}}{a}$

14. $\int xe^{ax} \, dx = \dfrac{e^{ax}}{a^2} (ax - 1)$

15. $\int e^{ax} \ln x \, dx = \dfrac{e^{ax} \ln x}{a} - \dfrac{1}{a} \int \dfrac{e^{ax}}{x} \, dx$

16. $\int \dfrac{dx}{1 + e^x} = x - \ln (1 + e^x) = \ln \left(\dfrac{e^x}{1 + e^x} \right)$

17. $\int (a^x - a^{-x}) \, dx = \dfrac{a^x + a^{-x}}{\ln a}$

18. $\int \dfrac{xe^{ax}}{(1 + ax)^2} \, dx = \dfrac{e^{ax}}{a^2 (1 + ax)}$

19. $\displaystyle\int xe^{-x^2}\, dx = -\frac{1}{2}\, e^{-x^2}$

20. $\displaystyle\int e^{ax}\, [\cos\, (bx)]\, dx = \frac{e^{ax}}{a^2 + b^2}\, [a\, \cos\, (bx) + b\, \sin\, (bx)]$

Table A5. Taylor's series formulas.

$$e^x = 1 + \frac{x}{1!} + \frac{x^2}{2!} + \frac{x^3}{3!} + \ldots, \text{ for all } x$$

$$a^x = 1 + \frac{x \ln a}{1!} + \frac{(x \ln a)^2}{2!} + \frac{(x \ln a)^3}{3!} + \ldots, \text{ for all } x$$

$$\log x = 2\left[\frac{x - 1}{x + 1} + \frac{1}{3}\left(\frac{x - 1}{x + 1}\right)^3 + \frac{1}{5}\left(\frac{x - 1}{x + 1}\right)^5 + \ldots\right], \text{ for } x > 0$$

$$\log\,(1 + x) = x - \frac{x^2}{2} + \frac{x^3}{3} - \frac{x^4}{4} + \frac{x^5}{5} - \ldots, \text{ for } -1 < x \le 1$$

$$\sin x = x - \frac{x^3}{3!} + \frac{x^5}{5!} - \frac{x^7}{7!} + \ldots, \text{ for all } x$$

$$\cos x = 1 - \frac{x^2}{2!} + \frac{x^4}{4!} - \frac{x^6}{6!} + \ldots, \text{ for all } x$$

$$\tan x = x + \frac{1}{3}\, x^3 + \frac{2}{15}\, x^5 + \frac{17}{315}\, x^7 + \ldots, \text{ for } 0 < |x| < \pi$$

$$\cot x = \frac{1}{x} - \frac{1}{3}\, x - \frac{1}{45}\, x^3 - \frac{2}{945}\, x^5 - \ldots, \text{ for } 0 < |x| < \pi$$

$$\arcsin x = x + \frac{x^3}{6} + \frac{3x^5}{40} + \frac{15x^7}{336} + \ldots, \text{ for } |x| \le 1$$

$$\arccos x = \frac{\pi}{2} - \arcsin x, \text{ for } |x| \le 1$$

$$\arctan x = x - \frac{x^3}{3} + \frac{x^5}{5} - \frac{x^7}{7} + \frac{x^9}{9} - \ldots, \text{ for } |x| \le 1$$

$$\text{arc cot } x = \frac{\pi}{2} - \arctan x, \text{ for } |x| \le 1$$

(continued)

$$\sinh x = x + \frac{x^3}{3!} + \frac{x^5}{5!} + \frac{x^7}{7!} + \frac{x^9}{9!} - \ldots, \text{ for all } x$$

$$\cosh x = 1 + \frac{x^2}{2!} + \frac{x^4}{4!} + \frac{x^6}{6!} + \frac{x^8}{8!} + \ldots, \text{ for all } x$$

$$\tanh x = x - \frac{1}{3} x^3 + \frac{2}{15} x^5 - \frac{17}{315} x^7 + \ldots, \text{ for } |x| < \frac{\pi}{2}$$

$$\coth x = \frac{1}{x} + \frac{1}{3} x - \frac{1}{45} x^3 + \frac{2}{945} x^5 - \ldots, \text{ for } 0 < |x| < \pi$$

Part B

Statistics Formulas and
Data Regression

ANALYSIS OF VARIANCE

The one-way analysis of variance tests the differences between the population means of k treatment groups. Group i ($i = 1, 2, \ldots, k$) has n_i observations (treatment group may have equal or unequal number of observations).

$Sum_i = $ sum of observations in treatment group i

$$= \sum_{j=1}^{n_i} x_{ij}$$

$$Total\ SS = \sum_{i=1}^{k} \sum_{j=1}^{n_i} x_{ij}^2 - \frac{\left(\sum_{i=1}^{k} \sum_{j=1}^{n_i} x_{ij} \right)^2}{\sum_{i=1}^{k} n_i}$$

$$Treat\ SS = \sum_{i=1}^{k} \frac{\left(\sum_{j=1}^{n_i} x_{ij} \right)^2}{n_i} - \frac{\left(\sum_{i=1}^{k} \sum_{j=1}^{n_i} x_{ij} \right)^2}{\sum_{i=1}^{k} n_i}$$

$$Error\ SS = Total\ SS - Treat\ SS$$

$$df_1 = Treat\ df = k - 1$$

$$df_2 = Error\ df = \sum_{i=1}^{k} n_i - k$$

$$Treat\ MS = \frac{Treat\ SS}{Treat\ df}$$

$$Error\ MS = \frac{Error\ SS}{Error\ df}$$

$$F = \frac{Treat\ MS}{Error\ MS} \left(\text{with } k - 1 \text{ and } \sum_{i=1}^{k} n_i - k \text{ degrees of freedom} \right)$$

THE STEP-WISE PROCEDURE FOR ONE-WAY ANALYSIS OF VARIANCE

- Select the risk of asserting that the factor effect exists when it actually does not (i.e., the α-risk).
- Tabulate the response data, Y_{ij}, for $i = 1, \ldots, k$ factor levels. Include in this tabulation the number of replicates for each factor level, n_i. For each factor level compute the following:

Response total

$$T_i = \sum_{j=1}^{n_i} Y_{ij}$$

Response average

$$\bar{Y}_i = T_i/n_i$$

Sum of squared responses

$$\sum_{j=1}^{n_i} Y_{ij}^2$$

- Compute the following by summing across the k factor levels:

Grand total

$$T = \sum_{i=1}^{k} T_i$$

Total experiments

$$N = \sum_{i=1}^{k} n_i$$

Total sum of squared responses

$$\sum_{i=1}^{k} \sum_{j=1}^{n_i} Y_{ij}^2 = \sum_{ij} Y_{ij}^2$$

- Compute the quantities:

$$\sum_{i=1}^{k} \left(\frac{T_i^2}{n_i} \right) \text{ and } \frac{T^2}{N} \text{ (Global Average)}$$

- Compute the *error sum of squares:*

$$SSE = \sum_{ij} Y_{ij}^2 - \sum_{i} \left(\frac{T_i^2}{n_i} \right) = \sum_{ij} (Y_{ij} - \bar{Y}_i)^2$$

This quantity is essentially a measure of the experimental error.

- Compute the *factor sum of squares:*

$$SSA = \sum_{i} \left(\frac{T_i^2}{n_i} \right) - \frac{T^2}{N} = \sum_{i} n_i (\bar{Y}_i - \bar{\bar{Y}})^2$$

This quantity reflects the experimental error plus the factor effect.

- Compute the *mean squares,* which is simply the sum of squares divided by the degrees of freedom, i.e.:

$$MSE = SSE/\nu_E, \text{ where } \nu_E = N - k$$

$$MSA = SSA/\nu_A, \text{ where } \nu_A = k - 1$$

- Compute the F^* statistic as follows:

$$F^* = MSA/MSE$$

- From Table B1 (p. 166) obtain the F-value for $100 (1 - a)\%$ confidence ν_A, ν_E degrees of freedom.
- The following criteria are applied to draw a conclusion from the analysis:

  - If $F^* > F$, we may assert that the factor effect exists with $100 (1 - \alpha)\%$ confidence.
  - If $F^* < F$, we conclude that there is insufficient evidence to support the assertion that the factor effect exists.

The following problem demonstrates application.

Example:

In the manufacture of plastic pipe, the melted plastic is extruded through an annular space between a die body and a mandrel. As the pipe is withdrawn from the die it is air cooled and cut into specified lengths. In addition to the length, the pipe must meet manufacturing specifications for wall thickness (minimum and maximum). The perfect part would naturally have uniform thickness and hence be perfectly concentric. Extruder variables that can affect manufacturing are: out of roundness of the die and mandrel and machine surging. It is important therefore, to have a quantitative understanding of how the operating variability of the equipment impact on the quality of the extruded part.

The study evaluates the degree of control in such a plant by obtaining wall thickness measurements of extruded pipe made from two similar machines [a newly conditioned extruder (Extruder A) and an older unit of the same design (Extruder B)]. Ten samples of pipe from each extruder were obtained, and wall thickness was measured at eight evenly spaced locations moving clockwise around the circumference.

Solution:

The data was tabulated on a Lotus 1-2-3 spreadsheet and is shown on the left-hand side (lhs) of Table B2 (p. 169).

The right-hand side (rhs) of Table B2 shows the calculations performed in the analysis. The sums of squared deviations (*ss*) are calculated as:

$$ss_{total} = \sigma^2(n - 1) = \Sigma\,(X_{i,j}^2) - n\bar{\bar{X}}^2$$

where $\bar{\bar{X}}$ = global average. It may also be computed by:

$$ss_{total} = \Sigma(X_{i,j}^2) - (\text{Global Total})^2/n$$

The row denoted "SS-Total" on the rhs of Table B2 is simply the sum of the rows minus the global mean:

$$ss_{total} = (-1)^2 + (2)^2 + \ldots + (-2)^2 - 80(-2.95)^2 = 767.8$$

Similarly, the *ss* attributable to differences in position (i.e., to the different points along the pipe's circumference) is computed from the averages of the eight points:

$$ss_{columns} = r\Sigma(\bar{X}_j^2) - n\bar{\bar{X}}^2$$

where

$\overline{X}_j$ = average for column j
r = number of rows (i.e., of points in a column)

That is, the column *ss* is the squared deviation attributed to the differences between the columns. Physically, the variation is due to measurements taken at different positions along the circumference. So, the calculation on the lhs of Table B2 is:

$$ss_{columns} = 10\ [(-1.8)^2 + (1.5)^2 + \ldots + (-3.5)^2] - 80(-2.95)^2]$$

$$= 513.4$$

The deviation within columns is defined as the residual:

$$SS_{residual} = SS_{total} - SS_{column}$$

The mean squared deviations are the variances (i.e., column and residual). The ratio between the column mean squared deviations and that of the residual constitutes an *F*-ratio. That is, the *F*-value is the ratio between the explained and unexplained variances. For a large *F*, the factor is significant (the level of significance is the probability value, *P*).

From the calculations in Table B2, Extruder A shows that the location of measurement along the circumference is important. We might therefore consider some die adjustments to make Extruder A's production more uniform.

Example:

We wish to construct a capacitance probe to be used in an experimental fluid bed to study bubble phenomenon (the change in bed capacitance reflects the presence of bubbles). Several probe geometries are being evaluated by immersing the probe in a static bed of fine particles. The objective of this analysis is to evaluate whether the sensor geometry influences the measurement. The capacitance measurements (picoFarads) obtained from four geometries are as follows.

Probe Configuration

	I	II	III	IV	
Capacitance (*pF*)	30	34	25	32	

(continued)

	I	II	III	IV	
Capacitance (pF)	36	41	26	29	
	42	47	25	35	
	29	50	29	43	
	40	51	33	41	
Total	177	223	138	180	718 = Grand Total
n_i	5	5	5	5	20 = N·
$\overline{Y}_i$	35.4	44.6	27.6	36.0	35.9 = $\overline{Y}$ (Global Average)
$\sum_j Y_{ij}^2$	6401	10147	3856	6620	27024 = $\sum_{ij} Y_{ij}^2$

$$\sum_i \frac{T_i^2}{n_i} = \frac{(177)^2}{5} + \frac{(223)^2}{5} + \frac{(138)^2}{5} + \frac{(180)^2}{5} = 26500.4$$

$$\frac{(\text{Grand Total})^2}{N} = \frac{(718)^2}{20} = 25776$$

Computed error sum of squares:

$$SSE = \sum_{ij} Y_{ij}^2 - \sum_i \frac{T_i^2}{n_i} = 27024 - 26500.4 = 523.6$$

Computed factor sum of squares:

$$SSA = \sum_i \frac{T_i^2}{n_i} - \frac{T^2}{N} = 26500.4 - 25776 = 724.4$$

Computed mean squares:

$$\nu_E = N - k = 20 - 4 = 16$$

$$MSE = \frac{SSE}{\nu_E} = \frac{523.6}{16} = 32.7$$

$$\nu_A = k - 1 = 4 - 1 = 3$$

$$MSA = \frac{SSA}{\nu_A} = \frac{724.4}{3} = 241.5$$

Computed F^*:

$$F^* = \frac{MSA}{MSE} = \frac{241.5}{32.7} = 7.38$$

From Table Bl for $\alpha = 0.05$, $F_{0.95}(3,16) = 3.24$.

$F^* > F$, therefore, we may conclude that the probe geometry influences the sensitivity of the measurement.

Solution:

We may summarize the important computations for this illustration of ANOVA as follows:

Source of Variations	Degrees of Freedom	SS	MS	F^*
Probe Geometry	3	724	241.3	7.38
Error	16	524	32.7	

$F_{0.95}(3,16) = 3.24$

TWO- AND THREE-WAY ANOVAs

ANOVA can be readily applied to separate out the effects of two or more variables. In principle, the number of variables that can be studied in a system is infinite. However, in practice, when the number of variables exceeds three, the computations become too cumbersome. The application of two-way ANOVA is demonstrated in the following illustration.

Example:

Continuing with the problem presented earlier we wish to separate the effects of the die-and-mandrel geometry and of machine surging.

Solution:

In terms of the tabulations presented in the last table, what we need is an additional SS calculation, except this time for the rows of data:

$$SS_{Rows} = c\Sigma(\overline{X_i^2}) - n\overline{\overline{X}}^2$$

where

$\overline{X}_i$ = row averages
c = number of columns (i.e., number of data points in a row)

This expression may also be written in terms of the row and global totals:

$$SS_{Rows} = [\Sigma(\text{Row Total})^2/c] - (\text{Global Total})^2/n$$

This calculation represents time in the data.

Hence, the calculations for Extruder A are:

$$SS_{Rows} = 8[(0.14)^2 + (0.1409)^2 + \ldots + (0.1419)^2] - 80(.14005)^2$$

$$= 7.08 \times 10^{-5}$$

$$SS_{Residual} = SS_{Total} - SS_{Columns} - SS_{Rows}$$

$$= 7.67 \times 10^{-4} - 5.13 \times 10^{-4} - 7.08 \times 10^{-5}$$

$$= 1.832 \times 10^{-4}$$

Similar calculations are carried out for Extruder B and we can summarize the two-way ANOVA for this problem (using tabulations from the table) as follows:

Extruder A

Source	SS	Degree of Freedom	MS	F	P
Total	7.67×10^{-4}	79	—	—	—
Position (Columns)	5.13×10^{-4}	7	7.3×10^{-5}	25.1	$\ll 0.001$
Time (Rows)	7.08×10^{-5}	9	7.87×10^{-6}	2.70	0.01
Residual	1.832×10^{-4}	63	2.91×10^{-6}	—	—

Extruder B

Source	SS	Degree of Freedom	MS	F	P
Total	5.519×10^{-3}	79	—	—	—
Position (Columns)	8.09×10^{-4}	7	1.156×10^{-4}	15.81	$\ll 0.001$
Time (Rows)	4.2493×10^{-3}	9	4.721×10^{-4}	64.6	$\ll 0.001$
Residual	4.607×10^{-4}	63	7.31×10^{-6}	—	—

The above values indicate that both position and time (i.e., surging) significantly contribute to the variance of both machines. For Extruder A, position is a greater source of unsteady operation, whereas for Extruder B, surging appears to be a greater source than position (although positioning or lack of centering the die-mandrel arrangement is certainly significant).

Example:

This example illustrates application of a three-way ANOVA to assessing the effects of process variables on gel formation in a polymer made via a solution polymerization process. Polymer samples were made under the following conditions:

Experiment 1 – Base case polymer was made (reactor lined out).
Experiment 2 – The base modifier (controlling the pH of the solution) was lowered.
Experiment 3 – Conditions returned to base case polymer and alkyl rate lowered.
Experiment 4 – Normal production requires quenching the reactor product to stop the reaction. In this experiment no quenching was done.

The polymers were then extruded into a thin film and the number of gel particles per square inch of film were counted. The results are reported in the form of coded data below, as 1-Relative Gel Count (where the relative count is the ratio of measured to the mean for the control polymer).

Run	Low Base	Low Alkyl	No Quench	Total
1	0.03	0.60	0.58	1.21
	−0.95	0.45	0.68	0.18

(continued)

Run	Low Base	Low Alkyl	No Quench	Total
2	−1.05	0.48	0.75	0.18
	−1.61	0.56	0.76	−0.29
Total	−3.58	2.09	2.77	1.28

The extrusion tests were performed twice, and two different lengths of film were analyzed for each run. The ANOVA table is prepared in the usual manner—i.e., sums of squares (SS) and mean squares (MS) are calculated, but here we have duplicates, and so we need an additional variance calculation. This is the SS for differences between pairs, which is termed SS_{Error}. In general, it is computed by averaging the replicates and summing the squared deviations from this average of the individual measurements. For this case of duplicates, a simplified expression is:

$$SS_{Error} = \tfrac{1}{2}[(x_{11} - x_{12})^2 + (x_{21} - x_{22})^2 + \ldots + (x_{n1} - x_{n2})^2]$$

Hence, $SS_{Error} = 0.30$.
And calculating the sums of squares:

$$SS_{Total} = (0.03)^2 + (-0.95)^2 + \ldots + (0.76)^2 - (1.28)^2/12 = 7.50$$

$$SS_{Base} = (0.30)^2 + (-0.95)^2 + (-1.05)^2$$

$$+ (-1.61)^2 - (1.28)^2/12 = 4.46$$

$$SS_{Alkyl} = 0.97$$

$$SS_{Quench} = 1.80$$

$$SS_{Residual} = 7.50 - 4.46 - 0.97 - 1.80 = 0.27$$

Solution:

The ANOVA summary is:

Source	SS	Degrees of Freedom	MS	F	F^+
Total	7.50	11	—	—	—
Base	4.46	3	1.49	10.64	29.8
Alkyl	0.97	3	0.32	2.29	6.4

Source	SS	Degrees of Freedom	MS	F	F⁺
Quench	1.80	3	0.60	4.29	12.0
Error⁺	0.30	6	0.05	–	–
Residual	0.27	2	0.14	–	–

Note that we have computed an additional F-value based on the MSE. The degrees of freedom for the MSE are simply the number of pairs. Both F-values arrive at the same conclusion, although the calculation based on error makes a stronger statement—namely, that the base modifier has a dominant role in gel formation. We may also conclude that quenching is important, and that the alkyl concentration has a lesser effect.

We may take this analysis a step further by asking the question, how does the response vary with the significant factor level? The following steps apply to this analysis:

- Tabulate the k means $(\overline{Y}_i)$ in ascending order.
- Enter the ANOVA table to take the MSE and ν_E.
- Compute the standard error of the mean:

$$S_{\bar{y}} = \sqrt{MSE/n}$$

where $n = N/k$.

- Use Table B3 (Studentized ranges for different α-values, p. 171). For the appropriate α-value use the row entry for the error degrees of freedom and obtain values of the Studentized ranges for $p = 2, 3, \ldots, k$.
- Multiply the ranges by $S_{\bar{y}}$ to obtain the least significant ranges (LSR).
- Compare the differences among means with the LSR's in the following manner:
  - Compare the extreme means with LSR for $p = k - 1$.
  - Compare means $k - 1$ apart with LSR for $k - 1$.
  - Compare up to the adjacent means using LSR for $p = 2$.

BARTLETT'S CHI-SQUARE STATISTIC[3]

$$\chi^2 = \frac{f \ln s^2 - \sum_{i=1}^{k} f_i \ln s_i^2}{1 + \dfrac{1}{3(k-1)} \left[\left(\sum_{i=1}^{k} \dfrac{1}{f_i} \right) - \dfrac{1}{f} \right]}$$

[3]Suggested reference—A. Hald, *Statistical Theory with Engineering Applications*, John Wiley and Sons (1960).

where

s_i^2 = sample variance of the ith sample
f_i = degrees of freedom associated with s_i^2
i = 1, 2, . . ., k
k = number of samples

$$s^2 = \frac{\sum_{i=1}^{k} f_i \, s_i^2}{f}$$

$$f = \sum_{i=1}^{k} f_i$$

This χ^2 has a chi-square distribution (approximately) with $k - 1$ degrees of freedom, which can be used to test the null hypothesis that $s_1^2, s_2^2, . . ., s_k^2$ are all estimates of the same population variance σ^2 (H_0: Each of $s_1^2, s_2^2, . . ., s_k^2$ is an estimate of σ^2).

BAYES' FORMULA[4]

Suppose $E_1, E_2, . . ., E_n$ are n mutually exclusive and exhaustive events, and A is an event for which the conditional probabilities, $P[A/E_i]$ of A given E_i, are known. If $P[E_i]$ are given, then the conditional probability $P[E_k/A]$ of any one event E_k given A is:

$$P[E_k/A] = \frac{P[E_k] \, P[A/E_k]}{\sum_{i=1}^{n} P[E_i] \, P[A/E_i]}$$

where k can be 1, 2, . . ., or n.

BEHRENS-FISHER STATISTIC[5]

Consider $\{x_1, x_2, . . ., x_{n_1}\}$ and $\{y_1, y_2, . . ., y_{n_2}\}$ to be independent random samples from two normal populations with means μ_1, μ_2 (unknown). If the

---

[4]Suggested reference—E. Parzen, *Modern Probability Theory and its Applications*, John Wiley and Sons (1960).

[5]Suggested reference—Fisher and Yates, *Statistical Tables for Biological, Agricultural and Medical Research*, Hafner Pub. Co. (1970).

variances σ_1^2, σ_2^2 cannot be assumed equal, then the Behrens-Fisher statistic

$$d = \frac{\bar{x} - \bar{y} - D}{\sqrt{\dfrac{s_1^2}{n_1} + \dfrac{s_2^2}{n_2}}}$$

is used instead of the t statistic to test the null hypothesis

$$H_0: \mu_1 - \mu_2 = D$$

Critical values of this test are tabulated in the Fisher-Yates tables for various values of n_1, n_2, α and θ, where α is the level of significance and

$$\theta = \tan^{-1}\left(\frac{s_1}{s_2}\sqrt{\frac{n_2}{n_1}}\right)$$

Note the following definitions:

$$\bar{x} = \frac{\Sigma x_i}{n_1}$$

$$\bar{y} = \frac{\Sigma y_i}{n_2}$$

$$s_1^2 = \frac{\Sigma x_i^2 - [(\Sigma x_i)^2/n_1]}{n_1 - 1}$$

$$s_2^2 = \frac{\Sigma y_i^2 - [(\Sigma y_i)^2/n_2]}{n_2 - 1}$$

Example:

$x = 79,\ 84,\ 108,\ 114,\ 120,\ 103,\ 122,\ 120$
$y = 91,\ 103,\ 90,\ 113,\ 108,\ 87,\ 100,\ 80,\ 99,\ 54$

Solution:

$H_0: \mu_1 = \mu_2\ (D = 0),\ n_1 = 8,\ n_2 = 10,\ \bar{x} = 106.25$

$s_1/\sqrt{n_1} = 5.88,\ d = 1.73,\ \theta = 47.88°$

BINOMIAL DISTRIBUTION[6]

The binomial density function for given p and n is:

$$f(x) = \binom{n}{x} p^x (1 - p)^{n-x}$$

where

n = a positive integer
$0 < p < 1$
$x = 0, 1, 2, \ldots, n$

$$f(x + 1) = \frac{p(n - x)}{(x + 1)(1 - p)} f(x)$$

$$(x = 0, 1, 2, \ldots, n - 1)$$

is used to find the cumulative distribution:

$$P(x) = \sum_{k=0}^{x} f(k)$$

Note that:

1. $f(0) = P(0)$
2. The mean m and the variance σ^2 are given by:

$$m = np$$

$$\sigma^2 = np (1 - p)$$

BISERIAL CORRELATION COEFFICIENT[7]
(refer to Figure B1)

The biserial correlation coefficient r_b is used where one variable Y is quantitatively measured while the other continuous variable X is artificially

---

[6]Suggested reference—E. Parzen, *Modern Probability Theory and its Applications*, John Wiley and Sons (1960).

[7]Suggested reference—B. Ostle, *Statistics in Research*, Iowa State University Press (1963).

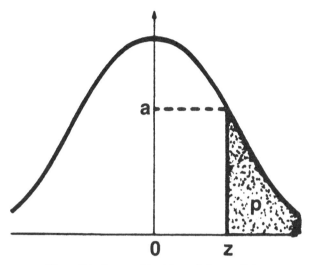

Figure B1. Illustrates biserial correlation coefficient.

dichotomized (that is, artificially defined by two groups). It measures the degree of linear association between X and Y:

$$r_b = \frac{n(\Sigma' y_i) - n_i \, \Sigma y_i}{na \, \sqrt{n \, \Sigma y_i^2 - (\Sigma y_i)^2}}$$

Suppose X takes the value 0 or 1.

Define:

n_1 = number of x's such that $x = 1$
n = total number of data points
$\Sigma' y_i$ = sum of the y's for which $x = 1$
Σy_i = sum of all y's
a = ordinate of the standard normal curve at point z cutting off a tail of that distribution with area equal to $p = n_1/n$.

This is illustrated in Figure B1. The necessary assumptions for a meaningful interpretation of r_b are:

1. Y is normally distributed.
2. The true distribution of X should be of normal form.

Example:

X_i	0	1	1	0	1	0	0	0	1
Y_i	3.1	2.8	5.6	0.3	2.5	2.4	4.8	2.9	7.7

Solution:

$$n_1 = 4,\ n = 9$$
$$a = 0.40,\ r_b = 0.59$$

BIVARIATE NORMAL DISTRIBUTION

This formula evaluates the joint probability density function:

$$f(x,y) = \frac{1}{2\pi\, \sigma_1\, \sigma_2\, \sqrt{1 - \varrho^2}}\ e^{-P(x,y)}$$

where:

$$P(x,y) = \frac{1}{2(1 - \varrho^2)} \left[\frac{(x - \mu_1)^2}{\sigma_1^2} - 2\varrho\, \frac{(x - \mu_1)(y - \mu_2)}{\sigma_1\sigma_2} + \frac{(y - \mu_2)^2}{\sigma_2^2} \right]$$

Note that:

1. $\sigma_1 \neq 0,\ \sigma_2 \neq 0$
2. $\varrho^2 < 1$

Example:

$$\mu_1 = -1,\ \sigma_1 = 1.5$$
$$\mu_2 = 1,\ \sigma_2 = 0.5$$
$$\varrho = 0.7$$

Solution:

$$f(1,2) = .04$$
$$f(-1,1) = .30$$

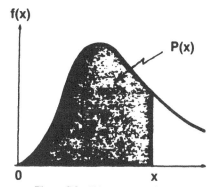

Figure B2. Chi-square distribution.

CHI-SQUARE DISTRIBUTION[8]

Given x, ν and $f(x)$, this test uses a series approximation to evaluate the chi-square cumulative distribution (see Figure B2):

$$P(x) = \int_0^x f(t)\, dt$$

$$= \frac{2x}{\nu} f(x)\left[1 + \sum_{k=1}^{\infty} \frac{x^k}{(\nu + 2)(\nu + 4) \ldots (\nu + 2k)} \right]$$

where
$x \geq 0$,
$\nu =$ the degrees of freedom, and
the density function is given by:

$$f(x) = \frac{x^{(\nu/2)-1}}{2^{(\nu/2)}\Gamma\left(\dfrac{\nu}{2}\right) e^{(x/2)}}$$

One can compute successive partial sums of the series. When two consecutive partial sums are equal, the value is used as the sum of the series.

[8]Suggested reference—Abramowitz and Stegun, *Handbook of Mathematical Functions*, National Bureau of Standards (1968).

CHI-SQUARE EVALUATION[9]

(EXPECTED VALUES EQUAL)

Test when the expected frequencies are equal.

$$\chi^2 = \sum_{i=1}^{n} \frac{(O_i - E_i)^2}{E_i} = \frac{n\Sigma O_i^2}{\Sigma O_i} - \Sigma O_i$$

where

O_i = observed frequency
E_i = expected frequency = $\Sigma O_i/n$

Example:

A die is tossed 120 times and the following frequencies are observed:

Number	1	2	3	4	5	6
Frequency, O_i	25	17	15	23	24	16

Solution:

$$\chi^2 = 5.00$$

(EXPECTED VALUES UNEQUAL)

$$\chi^2 = \sum_{i=1}^{n} \frac{(O_i - E_i)^2}{E_i}$$

where

O_i = observed frequency
E_i = expected frequency

The χ^2 statistic measures the closeness of the agreement between the observed frequencies and expected frequencies.

[9]Suggested reference—J. E. Freund and R. E. Walpole, *Mathematical Statistics*, Prentice-Hall (1962).

In order to apply the goodness of fit test to a set of given data, combining some classes may be necessary to make sure that each expected frequency is not too small (i.e., <5).

Example:

$O_i =$	8	50	47	56	5	14
$E_i =$	9.6	46.75	51.85	54.4	8.25	9.15

Solution:

$$\chi^2 = 4.84$$

COMBINATION

A combination is a selection of one or more of a set of distinct objects without regard to order. The number of possible combinations, each containing n objects, that can be formed from a collection of m distinct objects is given by:

$$_mC_n = \frac{m!}{(m - n)! \, n!} = \frac{m(m - 1) \ldots (m - n + 1)}{1 \cdot 2 \cdot \ldots \cdot n}$$

where m, n are integers and $0 \leq n \leq m$.

One can compute $_mC_n$ using the following algorithm:

1. If $n \leq m - n$,

$$_mC_n = \frac{m - n + 1}{1} \cdot \frac{m - n + 2}{2} \cdot \ldots \cdot \frac{m}{n}$$

2. If $n > m - n$, one computes $_mC_{m-n}$

Note that:

1. $_mC_n$, which is also called the binomial coefficient, can be denoted by:

$$C_n^m, \; C(m,n), \; \text{or} \; \binom{m}{n}$$

2. $_mC_n = {}_mC_{m-n}$
3. $_mC_0 = {}_mC_m = 1$
4. $_mC_1 = {}_mC_{m-1} = m$

CORRELATION COEFFICIENT[10]

Under the assumptions of normal correlation analysis, the following t statistic can be used to test the null hypothesis $\varrho = 0$,

$$t = \frac{r\sqrt{n-2}}{\sqrt{1-r^2}}$$

where r is an estimate (based on a sample of size n) of the true correlation coefficient ϱ. This t statistic has the t distribution with $n-2$ degrees of freedom.

To test the null hypothesis $\varrho = \varrho_0$, the z statistic is used.

$$z = \frac{\sqrt{n-3}}{2} \ln \frac{(1+r)(1-\varrho_0)}{(1-r)(1+\varrho_0)}$$

where z has approximately the standard normal distribution.

Example:

$$r = 0.12, n = 31$$

Solution:

In this case $t = 0.65$ and $z = 0.64$ (for $\varrho_0 = 0$)

COVARIANCE AND CORRELATION COEFFICIENT

For a given set of data points $\{(x_i, y_i), i = 1, 2, \ldots, n\}$, the covariance and the correlation coefficient are defined as:

$$\text{covariance } s_{xy} = \frac{1}{n-1}\left(\Sigma x_i y_i - \frac{1}{n}\Sigma x_i \Sigma y_i\right)$$

[10]Suggested references—Hogg and Craig, *Introduction to Mathematical Statistics*, Macmillan Co. (1970). J. Freund and R. W. Walpole, *Mathematical Statistics*, Prentice-Hall (1971).

or

$$s_{xy}' = \frac{1}{n}\left(\Sigma x_i y_i - \frac{1}{n}\Sigma x_i \Sigma y_i\right)$$

correlation coefficient $r = \dfrac{s_{xy}}{s_x s_y}$

where s_x and s_y are standard deviations:

$$s_x = \sqrt{\frac{\Sigma x_i^2 - (\Sigma x_i)^2/n}{n-1}}$$

$$s_y = \sqrt{\frac{\Sigma y_i^2 - (\Sigma y_i)^2/n}{n-1}}$$

Note:

$$-1 \le r \le 1$$

DIFFERENCES AMONG PROPORTIONS[11]

Suppose $x_1, x_2, \ldots, x_k$ are observed values of a set of independent random variables having binomial distributions with parameters n_i and θ_i ($i = 1, 2, \ldots, k$).

A chi-square statistic given by:

$$\chi^2 = \sum_{i=1}^{k} \frac{(x_i - n_i\hat{\theta})^2}{n_i\theta\,(1-\hat{\theta})}$$

can be used to test the null hypothesis $\theta_1 = \theta_2 = \ldots = \theta_k$, where:

$$\hat{\theta} = \sum_{i=1}^{k} x_i \bigg/ \sum_{i=1}^{k} n_i$$

[11]Suggested reference—J. Freund, *Mathematical Statistics*, Prentice-Hall (1971).

This χ^2 has the chi-square distribution with $k - 1$ degrees of freedom.

Example:

	N_i	X_i
Sample 1	400	232
Sample 2	500	260
Sample 3	400	197

Solution:

$$\chi^2 = 6.47 \qquad \theta = 0.53$$

DIMENSIONAL ANALYSIS[12]

Dimensional analysis or similarity theory is based on the principles of selecting a group of analogous phenomena from unrelated processes. For example, although the motion of fluids in the atmosphere and through piping are different systems, these flow phenomena are analogous because both represent the motion of viscous fluids under the influence of pressure gradients. Hence, the fluid motion for these different systems may be described by the unified Navier-Stokes equations and are considered to be of the same class of phenomena. Similarly, the motion of incompressible and compressible viscous fluids through piping and equipment, although considerably different processes, comprises a class of similar phenomena. Phenomena are considered operationally similar when the ratios of certain parameters characterizing the system are compatible.

We consider the condition used to determine operational similarity, namely the *geometric* condition. For a model to be geometrically similar to the prototype, the ratio of the distances between any two common points in both systems must be constant. This ratio is referred to as the *geometric scale factor*. For example, the characteristic dimensions of a drum dryer are its diameter, D, and length, L. The geometric scale factor is thus L/D, and the model will be similar to the prototype when the following condition is fulfilled:

$$\Lambda = \frac{L_1}{D_1} = \frac{L_2}{D_2}$$

---

[12]Suggested reference—D. S. Azbel and N. P. Cheremisinoff, *Fluid Mechanics and Unit Operations*, Ann Arbor Science, Ann Arbor, Michigan (1983).

where L and D have the same units (m) and thus, the ratio is dimensionless. Subscripts 1 and 2 refer to the model dryer and prototype, respectively. Thus, the model's diameter differs from the prototype by some constant scale factor. Similar ratios thus can be defined for any geometric configuration. Parameters of the same generic class are interchangeable; that is, those parameters that determine the similarity scale factors may be changed by similar values, which define the system geometry. Hence,

$$\frac{\ell'}{\ell''} = \frac{\ell_1'}{\ell_1''} = \frac{\ell_2'}{\ell_2''} = \frac{\ell_1' - \ell_2'}{\ell_1'' - \ell_2''} = \frac{d\ell'}{d\ell''} = \Lambda_\ell$$

Geometric similarity between systems is a necessary condition for similarity of physical phenomena. For physical similarity to exist, all important parameters influencing the phenomena must be similar. These parameters often change as functions of time and space in each system. Technological processes are similar only under conditions of mutual fulfillment of geometrical and time similarities over the fields of physical values, as well as similarities of initial and boundary conditions. Similarity among physical parameters such as velocity, acceleration, density, pressure, etc., may be defined in a manner analogous to geometric conditions:

$$
\begin{aligned}
\text{for velocity, } v_1 &= \Lambda_v v_2 \\
\text{for acceleration, } a &= \Lambda_a a_2 \\
\text{for density, } \varrho_1 &= \Lambda_\varrho \varrho_2 \\
\text{for pressure, } P_1 &= \Lambda_p P_2 \\
\text{for time, } t_1 &= \Lambda_t t_2
\end{aligned}
$$

The similarity scale factors Λ_ℓ, Λ_v, Λ_ϱ, Λ_p . . . are constant for different compatible points of two similar systems but change depending on the size ratio of the prototype to the model.

Process similarity between the prototype and the model may be determined through the use of *characteristic parameters*. These parameters are expressed in the form of compatible ratios in the limit of each system and are illustrated in Figure B3. We shall denote a "dummy" characteristic parameter by i.

$$i = \frac{\ell_1'}{L'} = \frac{\ell_1''}{L''} = \text{inv.}$$

where inv. denotes invariantly or "one and the same."

A parameter may be expressed in terms of relative units and hence, an appropriate scale can be defined. For example, the diameter of the dryer may

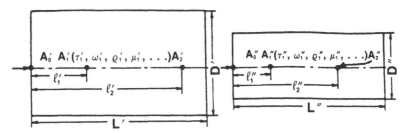

Figure B3. Scheme for the formulation of similarity conditions.

be selected as a scale rather than a length. For other systems, it may be appropriate to define this scale relative to common points in the model and prototype. Hence, from above we write:

$$\left. \begin{aligned} \frac{t_1'}{T'} &= \frac{t_1''}{T''} = i_t \\[6pt] \frac{v_1'}{v_0'} &= \frac{v_1''}{v_0''} = i_v \\[6pt] \frac{\varrho_1'}{\varrho_0'} &= \frac{\varrho_1''}{\varrho_0''} = i_\varrho \\[6pt] \frac{P_1'}{P_0'} &= \frac{P_1''}{P_0''} = i_p \end{aligned} \right\}$$

The characteristic parameters, i_t, i_v, i_ϱ, i_p . . . (referred to as simple) may not be equal to each other for different compatible points of similar systems, and are independent of the size ratio of the prototype and model. This means that in passing from one system to the other the characteristic parameters do not change. For example, assume a portion of fluid flows from a section I-I (a distance of 2 m from the inlet of a dryer) to the section II-II (a distance of 5 m from the inlet) for a total dryer length of $L = 10$ m. The characteristic parameter changes from $i_1 = 2/10 = 0.2$ to $i_2 = 5/10 = 0.5$; however, the scale factor remains unchanged if the sizes of the prototype and model are kept constant.

The mathematical principles in similarity theory are simple and easily implemented. However, the student is cautioned against adopting a formal or stereotyped approach to avoid direct errors that can result if the physical concepts of the methods are not adopted. The application of specific techniques or theorems is determined by a volume of preliminary knowledge about the process under consideration. Application of the theory to the development of fully integrated equations is known as dimensional analysis. There are three basic theorems, each of which addresses a separate issue concerning the planning of model experiments. These issues are:

1. Identification of important parameters to be measured in the experiments
2. The final and most useful form in which the experimental results should appear
3. The type of equipment to which the model's experimental results may be applied

NEWTON'S THEOREM

This addresses the first issue:

1. According to Newton, *the conditions necessary (and sufficient) for similarity of the phenomena are equality of the values of the dimensionless groups made up of the quantities given in the conditions.*
2. According to Kirpichev (1953) [see also Gukhman (1965) and Sedov (1959)], *the conditions necessary (and sufficient) for similarity of the phenomena are equality of characteristic parameters (similarity indicators) to unity.*

We will illustrate the correctness of these formulations in the following example. Consider two similar systems that satisfy Newton's second law (the momentum equation), determining the relation between external forces and produced acceleration.

From Newton's law, the total force acting on a body is:

$$f = m \frac{dv}{dt}$$

where

f = force (N)
m = mass (kg)
v = velocity (m/s)
t = time (s)

For two similar systems, the force expression may be written twice:

$$f_1 = m_1 \frac{dv_1}{dt_1}$$

$$f_2 = m_2 \frac{dv_2}{dt_2}$$

Physical values for the two systems differ only by a scale factor, therefore:

$$f_1 = \Lambda_f f_2$$
$$m_1 = \Lambda_m m_2$$
$$v_1 = \Lambda_v v_2$$
$$t_1 = \Lambda_t t_2$$
$$dv_1 = \Lambda_v dv_2$$
$$dt_1 = \Lambda_t dt_2$$

Dividing the equations for f_1 and f_2 we obtain:

$$\frac{f_1}{f_2} = \frac{m_1}{m_2} \frac{dv_1}{dv_2} \frac{dt_2}{dt_1} \text{ or } \Lambda_f = \Lambda_m \frac{\Lambda_v}{\Lambda_t}$$

or

$$\frac{\Lambda_f \Lambda_t}{\Lambda_m \Lambda_v} = 1 = \tilde{j}$$

According to Kirpichev's statement, the nondimensional parameter $\tilde{j}$ (indicator of similar transformation) for two similar phenomena is equal to unity. Therefore, the selection of numerical values of scale factors is not arbitrary but rather subjected to the conditions of $\tilde{j} = 1$.

$$\frac{f_1 dt_1}{m_1 dv_1} = \frac{f_2 dt_2}{m_2 dv_2} \text{ and } \frac{f_1 t_1}{m_1 v_1} = \frac{f_2 t_2}{m_2 v_2}$$

This last expression is a dimensionless group known as the Newton number (*Ne*):

$$Ne = \frac{ft}{mv}$$

or, taking into account that $t = \ell/v$,

$$Ne \equiv \frac{f\ell}{mv^2}$$

The symbol "$\equiv$" indicates that this is a definition and does not denote the Newton number to be a function of the f, t, m, v, i.e., $Ne \neq f(ft/mv)$. Thus, for a series (groups) of similar processes whose class is described by initial physical equations based on Newton's second law, the following equality is correct:

$$Ne_1 = Ne_2 = Ne_3 = \ldots = \text{idem}$$

Hence, if we consider similar processes of motion in a model and in a prototype, then:

$$Ne_{model} = Ne_{prototype}$$

This equation is the mathematical formulation of the first theorem of similarity. Thus, with properly planned experiments it is only necessary to measure those values that appear in the dimensionless groups of a process under evaluation. The dimensionless groups constitute generalized characteristics of a process consisting of dimensional physical values reflecting different characteristics of a phenomenon.

Because the dimensions entering these groups are reduced, they have a zero dimension and the numerical value of a dimensional group remains true whatever the system of units in which the various quantities are measured.

To derive and compute dimensionless groups properly, we must be sure that the initial equation or expression is dimensionally homogeneous, i.e., that the dimensions of all the terms in each group are consistent and cancel. The correctness of deriving dimensionless groups is checked by reducing the dimensions from which the group is formed. In the case considered,

$$Ne = \frac{ft}{mv}\left[\frac{\text{N-s}^2}{\text{kg m}} = \frac{\text{N}}{\text{N}}\right]$$

i.e., the dimensions are reduced and, consequently, the dimensionless group is correct. The *dimensionless groups* are derived from the *dimensional* equation governing the physical process. The essential premise of deriving these groups is the availability of the equation governing the process, i.e., its mathematical description. Regardless how this equation is expressed—in algebraic or differential form—the dimensionless groups may be derived by the same technique from any homogeneous dimensional equation.

Dimensionless groups can be formed simply by dividing through the equation by any dimensional product that makes all the terms of the equation dimensionless. Thus, they can be formed directly from the basic equation of the system and *without its formal solution*.

In addition to the basic physical equation, the derived form of the dimensionless group has an intrinsic physical character. For example, the Newton number expresses the ratio of active to reactive forces and is therefore a measure of impulse, *ft*, and momentum, *mv*. In special cases of motion depending on concrete expressions for force, mass and velocity, the Newton number can take on another form, that is, it can be transformed into the Reynolds, Euler, Froude and/or Stokes numbers.

To derive these dimensionless numbers, it is necessary to substitute the acting forces into the expression of the Newton number accordingly.

FROUDE NUMBER

For the Froude number (*Fr*):

$$Ne_1 = \frac{[mg]\ell}{mv^2} = \frac{g\ell}{v^2}$$

$$Fr = \frac{v^2}{g\ell}$$

Fr is the ratio of inertia force on a fluid element to the gravity force.

REYNOLDS NUMBER

For deriving the Reynolds number (*Re*) from the Newton number, we write down the expression of the friction force in the flow of a viscous fluid as it is determined by Newton's law:

$$f = \mu\ell^2 \frac{dv}{d\ell}$$

Substituting for *f* and denoting $m = \varrho\ell^3$ (*m* is related to a unit volume), we obtain

$$Ne_2 = \frac{\left[\mu\ell^2 \dfrac{dv}{d\ell}\right]\ell}{[\varrho\ell^3]v^2}$$

$$Ne_2 d\ell = \frac{\mu dv}{\varrho v^2}$$

$$Ne_2 \int_0^{\ell} d\ell = \frac{\mu}{\varrho} \int_0^{v_{max}} v^{-2} dv$$

$$Ne_2 = \frac{\mu}{\varrho v \ell}$$

or

$$Re = \frac{dv\varrho}{\mu}$$

or

$$Re = \frac{dv\,\gamma}{\mu g}$$

The Reynolds number is interpreted as the ratio of inertial forces to the viscous forces in the flow.

EULER NUMBER

The Euler number (Eu) which characterizes the hydrodynamic processes running under action of mechanical pressure, is the ratio of static pressure drop, Δp, and dynamic head, ϱv^2:

$$Ne_3 = \frac{[\Delta p \ell^2]\ell}{[\varrho \ell^3]v^2}$$

$$Eu = \frac{\Delta p}{\varrho v^2}$$

STOKES NUMBER

The Stokes number (Stk) is important in analyzing sedimentation processes. Substituting the resistance force of the medium into the Newton number,

$$R = 3\pi d\mu v_1$$

$$Ne_4 = \frac{[3\pi d\mu v]\ell}{\left[\dfrac{\pi d^3}{6}(\varrho_1 - \varrho_2)\right]v_2^2} = \frac{\mu\ell}{d^2\varrho_1 v^2}$$

or

$$Stk = \frac{d^2\varrho_1 v_2}{\mu\ell}$$

The Stokes number is the ratio of the resistance force of the medium to the buoyancy force of a particle. The intrinsic physical character of each dimensionless number (criterion of similarity) differs from *arbitrarily chosen* dimensionless complexes composed of random physical values.

DERIVATION OF DIMENSIONLESS GROUPS FROM PROCESS-GOVERNING EQUATIONS
(see Table B4, p. 174)

There are three basic methods for deriving dimensionless groups from a governing equation describing the process. Each is illustrated by considering the steady accelerated motion of a body:

$$w = w_0 + at$$

where

w = velocity at time t (m/s)
w_0 = velocity at time $t = 0$ (m/s)
a = acceleration (m/s²)
t = time from starting motion (s)

METHOD I

This involves variable transformation through the use of scale factors used earlier in deriving the Newton number.
For the first phenomenon,

$$w_1 = w_{01} + a_1 t_1$$

For the second phenomenon,

$$w_2 = w_{02} + a_2 t_2$$

For similarity of two phenomena we have

$$\frac{w_1}{w_2} = C_w$$

$$w_1 = C_w w_2$$

$$\frac{w_{01}}{w_{02}} = c_w$$

$$w_{01} = c_w w_{02}$$

$$\frac{a_1}{a_2} = C_a$$

$$a_1 = C_a a_2$$

$$\frac{t_1}{t_2} = C_t$$

$$t_1 = C_t t_2$$

Substituting the new notations of w_1, w_{01}, a_1 and t_1 into the expression for the first phenomenon

$$C_w w_2 = c_w w_{02} + C_a C_t a_2 t_2$$

These equations can coexist only under the condition of reduction of multiples formed from factors. This is equivalent to the condition of equality in pairs:

$$C_w = c_w$$

$$C_w = C_a C_t$$

The last condition gives two *characteristic parameters* (indicators of similarity):

$$\tilde{j}' = \frac{C_w}{c_w} = 1$$

$$\tilde{j}'' = \frac{C_a C_t}{C_w} = 1$$

In the expression for $\tilde{j}'$, C_w is related to *different* process velocities but it cannot be reduced.

Substituting the scale factors by the ratios of values, we obtain:

$$\tilde{j}' = \frac{\dfrac{w_1}{w_2}}{\dfrac{w_{01}}{w_{02}}} = 1$$

$$\frac{w_1}{w_{01}} = \frac{w_2}{w_{02}} = \text{idem}$$

Hence, the first dimensionless number, a *dimensionless velocity* (in this case a relative velocity) is:

$$K_1 \equiv \frac{w}{w_0}$$

In the same manner we obtain from the expression $\tilde{j}''$ the second dimensionless number:

$$\tilde{j}'' = \frac{\dfrac{a_1}{a_2} \times \dfrac{t_1}{t_2}}{\dfrac{w_{01}}{w_{02}}} = 1$$

$$\frac{a_1 t_1}{w_{01}} = \frac{a_2 t_2}{w_{02}} = \text{idem}$$

Hence, the second dimensionless number of time similarity (i.e., the criterion of kinematic Homochronity) is:

$$K_2 \equiv \frac{at}{w_0}$$

A constant-scale value, w_0, is assumed in both expressions for the dimensionless numbers. According to the terms of the initial equation, we have to determine:

$$\frac{w}{w_0} = f\left(\frac{at}{w_0}\right)$$

or

$$K_1 = f(K_2)$$

In this simple case, the form of this function is understood in the second method of similarity transformation.

METHOD II

This method consists of dividing all terms of the homogeneous equation by one of its terms serving as a scale (in this case by w_0):

$$\frac{w}{w_0} = 1 + \frac{at}{w_0}$$

Hence,

$$K_1 = \frac{w}{w_0}, K_2 = \frac{at}{w_0}$$

whence the form of the function is evident:

$$K_1 + 1 + K_2$$

The equations governing the process are usually much more complicated and the form of the function not so readily determined. However, for clarification of the combination of physical values entering into the dimensionless numbers, the method is applicable.

METHOD III

This method consists of transition to new independent units of measurement of physical values. Assume we know a priori that there is an explicit relationship:

$$w = f(w_0, a, t)$$

Here, the base units of measure are m and s, and the dependent units of measure are m/s (for velocity) and m/s² (for acceleration). Now we transfer to new independent units of measure, which are less in L and T times than the first ones. Then the numerical values of w, a and t will be changed to:

$$w \frac{L}{T} = f\left(w_0 \frac{L}{T}, \ a \frac{L}{T^2}, \ tT \right)$$

Fluid motion is independent of the units chosen for measuring specific process characteristics. Therefore, the basic equation has to retain its structure at different values of coefficients L and T. The numerical values of these coefficients are so chosen to provide expressions that are simple and convenient in application. In particular, we may choose L and T as:

$$w_0 \frac{L}{T} = 1$$

$$a = \frac{L}{T^2} = 1$$

$$tT = 1$$

$$T = \frac{1}{t}$$

$$\frac{L}{T} = \frac{1}{w_0}$$

$$\frac{L}{T^2} = \frac{1}{a}$$

$$\frac{1}{w_0} = \frac{T}{a} = \frac{1}{at}$$

$$a \frac{L}{T} = a \frac{1}{w_0} \times \frac{1}{T} = \frac{at}{w_0}$$

Substituting the obtained values into the constitutive equation, we have

$$\frac{w}{w_0} = f\left(1, \ \frac{at}{w_0}, \ 1 \right)$$

or

$$\frac{w}{w_0} = f\left(\frac{at}{w_0} \right)$$

or

$$K_1 = f(K_2)$$

This coincides with the result of transformation by the first method.

All these methods are applicable if one has the equation governing the physical process, i.e., the mathematical description of the process containing all necessary characteristic physical values.

However, as is more often the case, only the *qualitative* description of a process is known; the quantitative relationships among the different physical factors and their form are unknown. Then the theory of similarity helps to determine the *hypothetical* form and prioritizes the most important dimensionless numbers of the process. We then may use carefully planned experiments to validate this hypothesis. As noted earlier, the method used to predict the form of a dimensionless number without knowledge of the basic equation is called dimensional analysis. Its application is conjugated with Buckingham's "pi" theorem.

The number of independent dimensionless products, whose values determine the properties of the system, is generally less than the number of different kinds of physical quantities in the system. The consequent reduction in the number of experimental conditions that need to be covered in a program of experiments represents an enormous simplification and economy, resulting in simplifications in the plotting or tabulation of results. Similar simplifications are achieved in analytical solutions when problems are formulated in terms of dimensionless products at the outset.

FEDERMAN-BUCKINGHAM'S THEOREM

Federman-Buckingham's theorem constitutes the second theorem of similarity. It states that the quantitative results of experiments should be presented by equations expressing the relationship among nondimensional numbers. The dimensionless groups, K_1, containing the parameters of interest should be expressed as a function of other dimensionless numbers reflecting different sides of a process:

$$K_1 = f(K_2, K_3, K_4, \ldots)$$

In the examples given above, the form of the function was known.

Generally, the form of this function is not known a priori and must be determined from experimental data. Unlike human beings, nature rigorously applies the laws of geometric progression and probability. Man attempts to approximate these laws through logarithmic relationships. The results of experimental investigations are approximated more readily in either a power form:

$$K_1 = CK_2^m K_3^n K_4^p$$

or exponential (for kinetic processes) forms:

$$K_1 = e^{-t/\theta} \text{ (damping process)}$$
$$K_1 = K_0(1 - e^{-t/\theta}) \text{ (growing process)}$$

where C, m, n and p are constants determined from the graphical analysis of experimental data.

K_0 is the initial value of the dimensionless number K_1 at time $t = 0$, and t denotes the time from the beginning of the process. θ is the time constant of the process that depends on the conditions of its realization and is expressed through the dimensionless numbers.

The second theorem of similarity is formally stated as follows: *The solution of any differential equation may be presented as a relationship among the dimensionless numbers obtained from this equation.* Analytical methods provide the initial description of phenomena in a form of complicated differential equations that determine interrelations among values in formulating a problem. However, their solution to design relationships usually is not achieved because of the complexity of the problem. Therefore, although a purely analytical investigation remains, which, as a rule, is the most desirable approach to problem solving, it often is not applied to engineering solutions. On the other hand, purely experimental approaches without knowledge of at least initial theoretical expressions often are doomed to failure. Blind or brute force experimentation often results in a tremendous volume of data, much of which is extraneous and unrelated to the desired solution.

Similarity theory brings to the *experimental* solution of the problem *physical laws* in the form of initial equations describing the process. The transition to generalized variables and modeling significantly facilitates and accelerates this solution.

Any dimensionless expression (despite the empirical method of its derivation in an explicit form) has a definite physical meaning because it reflects the laws of nature expressed by the initial system of physical equations. Modeling based on dimensional analysis produces only approximate solutions because only the most important parameters describing the phenomenon are included in the final expressions. Each dimensionless group in the generalized equation reflects some of the aspects of the process, and the total equation attempts to approximate the behavior of the process as a whole. The technique of treating experimental data in terms of power law expressions is illustrated in the following example.

Example:

A process may be described by a relationship in which $\pi = 2$. Develop a relationship for the function $K_1 = f(K_2)$.

Solution:

We may assume a general form of the relationship to be $K_1 = CK_2^n$, where C and n are the unknowns. Taking the logarithm of this expression, we obtain:

$$\log K_1 = \log C + n \log K_2$$

This is the equation of a straight line on log–log coordinates. If we plot the experimental data and it can be correlated by a straight line, then the slope will determine the coefficient n, and C can be computed for any point on the line. If the data cannot be correlated by a straight line, then there is a portion of the phenomenon not accounted for and further clarification of the system's physics is needed.

ERROR FUNCTION AND COMPLEMENTARY ERROR FUNCTION[13]

The error function is defined as:

$$\text{erf } x = \frac{2}{\sqrt{\pi}} \int_0^x e^{-t^2} \, dt$$

---

[13]Suggested reference — Abramowitz and Stegun, *Handbook of Mathematical Functions*, National Bureau of Standards (1968).

$$= \frac{2}{\sqrt{\pi}} e^{-x^2} \sum_{n=0}^{\infty} \frac{2^n}{1 \cdot 3 \cdot \ldots \cdot (2n + 1)} x^{2n+1}$$

The complementary error function is defined as:

$$\text{erfc } x = 1 - \text{erf } x$$

where $x > 0$.

EXPONENTIAL CURVE FIT[14]

For n pairs of data points $\{(x_i, y_i), i = 1, 2, \ldots, n\}$, where $y_i > 0$, an exponential function is of the form

$$y = a \, e^{bx} \ (a > 0)$$

This equation can be linearized into

$$\ln y = \ln a + bx$$

The following statistics are computed:

1. Coefficients a, b:

$$b = \frac{\Sigma x_i \ln y_i - \frac{1}{n}(\Sigma x_i)(\Sigma \ln y_i)}{\Sigma x_i^2 - \frac{1}{n}(\Sigma x_i)^2}$$

$$a = \exp\left[\frac{\Sigma \ln y_i}{n} - b \, \frac{\Sigma x_i}{n}\right]$$

2. Coefficient of determination:

$$r^2 = \frac{\left[\Sigma x_i \ln y_i - \frac{1}{n} \Sigma x_i \Sigma \ln y_i\right]^2}{\left[\Sigma x_i^2 - \frac{(\Sigma x_i)^2}{n}\right]\left[\Sigma(\ln y_i)^2 - \frac{(\Sigma \ln y_i)^2}{n}\right]}$$

[14]Suggested reference—A. Brownlee, *Statistical Theory and Methodology in Science and Engineering*, John Wiley and Sons (1965).

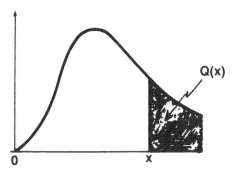

Figure B4. *F* distribution.

3. Estimated value $\hat{y}$ for a given x:

$$\hat{y} = a\ e^{bx}$$

Note that n is a positive integer and $n \neq 1$.

F DISTRIBUTION[15]

The integral of the *F* distribution:

$$Q(x) = \int_x^\infty \frac{\Gamma\left(\frac{\nu_1 + \nu_2}{2}\right) y^{(\nu_1/2)-1} \left(\frac{\nu_1}{\nu_2}\right)^{\nu_1/2}}{\Gamma\left(\frac{\nu_1}{2}\right)\Gamma\left(\frac{\nu_2}{2}\right)\left(1 + \frac{\nu_1}{\nu_2}\, y\right)^{(\nu_1+\nu_2)/2}}\ dy$$

for given values of x (>0), degrees of freedoms ν_1, ν_2, provided either ν_1 or ν_2 is even (see Figure B4).

The integral is evaluated by means of the following series:

1. ν_1 even

$$Q(x) = t^{(\nu_2/2)} \left[1 + \frac{\nu_2}{2}\,(1 - t) + \ldots \right.$$

$$\left. + \frac{\nu_2(\nu_2 + 2)\ \ldots\ (\nu_2 + \nu_1 - 4)}{2\cdot4\cdot\ \ldots\ \cdot(\nu_2 - 2)}\,(1 - t)^{(\nu_1-2)/2} \right]$$

[15]Suggested reference—Abramowitz and Stegun, *Handbook of Mathematical Functions*, National Bureau of Standards (1968).

$$Q(x) = 1 - (1 - t)^{(\nu_1/2)} \left[1 + \frac{\nu_1}{2} t + \ldots \right.$$

$$\left. + \frac{\nu_1(\nu_1 + 2) \ldots (\nu_2 + \nu_1 - 4)}{2 \cdot 4 \cdot \ldots \cdot (\nu_2 - 2)} t^{(\nu_2 - 2)/2} \right]$$

where

$$t = \frac{\nu_2}{\nu_2 + \nu_1 x}$$

Note that if using the smaller of ν_1, ν_2 could save computation time. For example, if $\nu_1 = 10$, $\nu_2 = 20$, then classify the problem as ν_1 is even to obtain the answer.

GAMMA FUNCTION[16]

The value of the gamma function $\Gamma(x)$:

$$\Gamma(x) = \int_0^\infty t^{x-1} e^{-t} dt$$

$$\cong \sqrt{2\pi/x} \; x^x e^{-[x - (1/12x) + (1/360x^3)]}$$

The figure below illustrates the function.

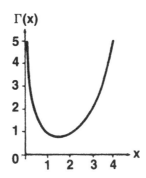

---

[16]Suggested reference—Abramowitz and Stegun. *Handbook of Mathematical Functions*, National Bureau of Standards (1968).

Suppose ϵ is the error, then:

$$\frac{\epsilon}{\Gamma(x)} < 2 \times 10^{-7}$$

This approximation is good for large x. In order to increase the accuracy (especially for small values of x), one can compute $\Gamma(x + 5)$, then $\Gamma(x)$ is calculated using the following formula:

$$\Gamma(x) = \frac{\Gamma(x + 5)}{(x + 4)(x + 3)(x + 2)(x + 1)x}$$

GENERALIZED MEAN

For a set of n positive numbers $\{a_1, a_2, \ldots, a_n\}$, the generalized mean is defined by:

$$M(t) = \left(\frac{1}{n} \sum_{k=1}^{n} a_k^t\right)^{1/t}$$

where t is any desired number.

Note:

1. If $t = 1$, the generalized mean $M(1)$ is the same as the arithmetic mean.
2. If $t = -1$, the generalized mean $M(-1)$ is the same as the harmonic mean.

GEOMETRIC MEAN

For a set of n positive numbers $\{a_1, a_2, \ldots, a_n\}$, the geometric mean is defined by:

$$G = \prod_{k=1}^{n} (a_k)^{1/n} = (a_1 \, a_2 \ldots a_n)^{1/n}$$

Example:

Given the data set: 2, 3.4, 3.41, 7, 11, 23

Solution:

Geometric Mean, $G = 5.87$

HARMONIC MEAN

For a set of n positive numbers $\{a_1, a_2, \ldots, a_n\}$, the harmonic mean is defined by:

$$H = \frac{n}{\dfrac{1}{a_1} + \dfrac{1}{a_2} + \ldots + \dfrac{1}{a_n}}$$

Example:

Given the data set: 2, 3.4, 3.41, 7, 11, 23

Solution:

Harmonic Mean, $H = 4.40$

HYPERGEOMETRIC DISTRIBUTION[17]

The hypergeometric density function for given a, b and n is:

$$f(x) = \frac{\dbinom{a}{x}\dbinom{b}{n-x}}{\dbinom{a+b}{n}}$$

where

a, b, n = positive integers
$\quad x \leq a$
$\quad n - x \leq b$
$\quad x = 0, 1, 2, \ldots, n$

---

[17]Suggested reference—J. E. Freund and R. E. Walpole, *Mathematical Statistics,* Prentice-Hall (1971).

The recursive relation:

$$f(x + 1) = \frac{(x - a)(x - n)}{(x + 1)(b - n + x + 1)} f(x)$$

$$(x = 0, 1, 2, \ldots, n - 1)$$

is used to find the cumulative distribution:

$$P(x) = \sum_{k=0}^{x} f(k)$$

Note that:

1. $f(0) = P(0)$
2. The mean m and the variance σ^2 are given by:

$$m = \frac{an}{a + b}$$

$$\sigma^2 = \frac{abn(a + b - n)}{(a + b)^2(a + b - 1)}$$

INVERSE NORMAL INTEGRAL[18]

This test determines the value of x such that:

$$Q = \int_{x}^{\infty} \frac{e^{-(t^2/2)}}{\sqrt{2\pi}} \, dt$$

where Q is given and $0 < Q \le 0.5$. See Figure B5.
The following rational approximation is used:

$$x = t - \frac{c_0 + c_1 t + c_2 t^2}{1 + d_1 t + d_2 t^2 + d_3 t^3} + \epsilon(Q)$$

[18]Suggested reference—Abramowitz and Steguń, *Handbook of Mathematical Functions*, National Bureau of Standards (1968).

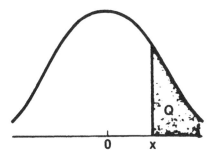

Figure B5. Inverse normal integral.

where

$$|\epsilon(Q)| < 4.5 \times 10^{-4}$$

$$t = \sqrt{\ln \frac{1}{Q^2}}$$

$$c_0 = 2.515517$$
$$c_1 = 0.802853$$
$$c_2 = 0.010328$$
$$d_1 = 1.432788$$
$$d_2 = 0.189269$$
$$d_3 = 0.001308$$

KENDALL'S COEFFICIENT OF CONCORDANCE[19]

Suppose n individuals are ranked from 1 to n according to some specified characteristic by k observers; the coefficient of concordance W measures the agreement between observers (or concordance between rankings):

$$W = \frac{12 \sum_{i=1}^{n} \left(\sum_{j=1}^{k} R_{ij} \right)^2}{k^2\, n(n^2 - 1)} - \frac{3(n + 1)}{n - 1}$$

[19]Suggested reference—J. D. Gibbons, *Nonparametric Statistical Inference*, McGraw-Hill (1971). M. G. Kendall, *Rank Correlation Methods*, Hafner Publishing Co. (1962).

where R_{ij} is the rank assigned to the ith individual by the jth observer.

W varies from 0 (no community of preference) to 1 (perfect agreement). The null hypothesis that the observers have no community of preference may be tested using special tables or, if $n > 7$, by computing:

$$\chi^2 = k(n - 1)W$$

which has approximately the chi-square distribution with $n - 1$ degrees of freedom.

Example:

Table for R_{ij} ($n = 10$, $k = 3$)

i \ j	1	2	3
1	6	7	3
2	1	4	2
3	9	3	5
4	2	6	1
5	10	8	9
6	3	2	6
7	5	9	8
8	4	1	4
9	8	10	10
10	7	5	7

$W = .69$
$\chi^2 = 18.64$

KRUSKALL-WALLIS STATISTIC[20]

This statistic enables one to test the null hypothesis that k independent random samples of sizes n_1, n_2, . . ., and n_k come from identical continuous populations.

Arrange all values from k samples jointly (as if they were one sample) in an increasing order of magnitude. Let R_{ij} ($i = 1, 2, . . ., k, j = 1, 2, . . ., n_i$) be the rank of the jth value in the ith sample.

[20]Alexander and Quade, *On the Kruskal-Wallis Three Sample H-Statistic,* University of North Carolina, Department of Biostatistics, Inst. Statistics Mimeo Ser. 602 (1968).

The Kruskal-Wallis statistic H can be used to test the null hypothesis.

$$H = \frac{12}{N(N+1)} \sum_{i=1}^{k} \frac{\left(\sum_{j=1}^{n_i} R_{ij}\right)^2}{n_i} - 3(N+1)$$

where

$$N = \sum_{i=1}^{k} n_i$$

When all sample sizes are large (>5), H is distributed approximately as chi-square with $k-1$ degrees of freedom. For small samples, the test is based on special tables.

Refer to Reference 21 for a table for small values (i.e., $k = 3$).

LINEAR REGRESSION
(METHOD OF LEAST SQUARES FIT)

The method of least squares can be used to arrive at the "best fitting curve" of a body of data. Figure B6 shows data points $(X_1, Y_1), (X_2, Y_2), \ldots, (X_N, Y_N)$. For a given value of X (say X_0), there will be a difference between the value Y_1 and corresponding value as determined by the curve. As shown in Figure B6 the difference (also called the *deviation, error,* or *residual*) is denoted by D_1 and may be positive, negative, or zero. Deviations for corresponding $X_2, \ldots, X_N$ values are $D_2, \ldots, D_N$.

An indication of how good the fit of the curve in Figure B6 is can be obtained from the sum of the squared deviations, i.e., $D_1^2 + D_2^2 + \ldots + D_N^2$. When this value is small, the fit is good, and conversely, when it is large, the fit is poor. Hence, the objective of the method of least squares is to find the equation of a curve such that

$$\min \{D_1^2 + D_2^2 + \ldots + D_N^2\}$$

The curve with this property is the *best fitting curve.* A line having this property is called a *least square line.*

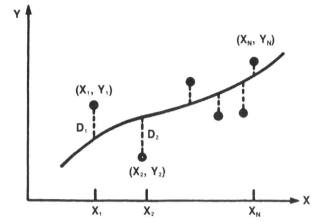

Figure B6. Illustrates regression and errors.

The general method used to arrive at the best fitting curve is referred to as linear regression analysis. The analysis estimates a function or model which will predict the behavior of a response variable Y over a range of factor levels X. We first treat the case of a single factor X, where a linear relationship between Y and X is sought. The relationship sought in the regression is illus-

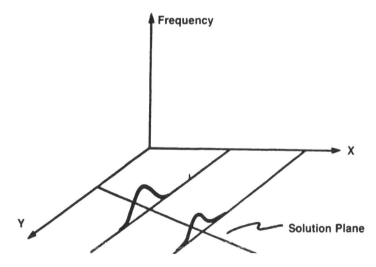

Figure B7. Conceptual example with confidence limits.

trated conceptually in Figure B7 in which we can model the solution approach as follows:

$$Y_{ij} = \beta_0 + \beta_1 X_1 + \epsilon_{ij} \tag{1}$$

where

X_i = the ith factor level
Y_{ij} = the response at the ith factor level and jth replication of the data
ϵ_{ij} = the error at the ith factor level and jth replication
β_0, β_1 = unknown parameters to be estimated from the regression (i.e., the intercept and slope of the linear relationship, respectively)

If no replicate experiments are run then the j subscript is omitted.

The following assumptions are applied in the regression scheme:

(1) X can be freely set to a value without error.
(2) The error term ϵ is normally distributed with zero mean and variance σ^2 independent of the value of X.

Parameters β_0 and β_1 are estimated by the least squares estimates b_0 and b_1 to obtain the following linear estimation equation:

$$\hat{Y} = b_0 + b_1 X \tag{2}$$

The estimates of b_0 and b_1 are those values which minimize:

$$\min \sum_{i=1}^{n} (Y_i - \hat{Y}_i)^2$$

The regression procedure is as follows:

• Compute the means $\overline{X}$ and $\overline{Y}$ and the deviations S_{xx}, S_{xy}, S_{yy} from the following formulas:

$$S_{xx} = \Sigma x^2 - (\Sigma x)^2/n \tag{3}$$

$$S_{yy} = \Sigma y^2 - (\Sigma y)^2/n \tag{4}$$

$$S_{xy} = \Sigma xy - (\Sigma x)(\Sigma y)/n \tag{5}$$

- Calculate the estimated slope of the linear equations, b_1 [for Equation (2)]:

$$b_1 = S_{xy}/S_{xx} \qquad (6)$$

- Compute the intercept estimate for Equation (2):

$$b_0 = \overline{Y} - b_1\overline{X} \qquad (7)$$

- Calculate the variance estimate:

$$S^2 = \frac{S_{yy} - b_1^2 S_{xx}}{n - 2} \qquad (8)$$

- Calculate the correlation coefficient:

$$r* = S_{xy}/\sqrt{S_{xx}\, S_{yy}} \qquad (9)$$

The above procedure provides us with a least squares prediction equation that passes through the point $(\overline{Y},\overline{X})$ and having a standard deviation of prediction S, and correlation coefficient $r*$.

Example:

A conductance probe is used to measure the liquid inventory in a holding tank. The probe has been calibrated by immersing it at different levels in the process fluid and measuring the output voltage. The calibration data, converted to gallons based on the tank's diameter are given below. Let's prepare a regression formula to obtain a direct reading of inventory during operation.

	X Output Voltage (mV)	Y Gallons
	0	0
	23	295
	36	405
	39	550
	67	850
	80	970
Total	245	3070
n	6	6
Average	40.8	511.7

Solution:

$$S_{xx} = 14{,}235 - (245)^2/6 = 4{,}230.8$$

$$S_{yy} = 2.217 \times 10^6 = (3070)^2/6 = 6.461 \times 10^5$$

$$S_{xy} = 1.774 \times 10^5 - (245)(3070)/6 = 52{,}007$$

$$b_1 = 52{,}007/4{,}230.8 = 12.29$$

$$b_0 = 511.7 - 12.29(40.8) = 10.17$$

$$S^2 = \frac{6.461 \times 10^5 - (12.29)^2(4{,}230.8)}{4} = 1765.7$$

$$S = 42.0$$

$$r^* = S_{xy}/\sqrt{S_{xx}\,S_{yy}} = 52{,}007/\sqrt{(4230.8)(6.461 \times 10^5)} = 0.9947$$

Hence, we have a regression equation:

$$\hat{Y} = 10.17 + 12.29X$$

or

$$\text{Gallons} = 10.17 + 12.29 \,(\text{Voltage})$$

Although the correlation coefficient is high (i.e., a good fit) we have a standard deviation for prediction of 42 gallons. When the tank is near full capacity this amounts to less than 5% error, but at low inventories, the predictive equation could be significantly off. For this reason, we need to quantify the confidence intervals of the regression model.

Confidence intervals can be established on β_1, $\beta_0 + \beta_1 X_0$ and Y. They can be computed for the following:

- β_1, to test the hypothesis, $\mu_0: \beta_1 = 0$.
- On the expected value of Y at any given X (e.g., X_0). This is essentially the confidence limit on the model.
- On future Y values at a given X (e.g., X_0). The procedure for calculating confidence intervals is as follows:
  - Select the α-risk. Use Student's t statistic for $n - 2$ degrees of freedom and $100\,(1 - \alpha)\%$ confidence.

— Compute the confidence interval for β_1 from:

$$\beta_1 = b_1 \pm ts/\sqrt{S_{xx}} \tag{10}$$

— Compute the confidence interval on the error:

$$\epsilon(Y) = \beta_0 + \beta_1 X \text{ at } X = X_0$$

using:

$$\beta_0 = b_0 + b_1 X_0 \pm ts \sqrt{\frac{1}{n} + \frac{(X_0 - \overline{X})^2}{S_{xx}}}$$

— Compute the confidence interval on Y at X_0 using:

$$Y = b_0 + b_1 X_0 \pm ts \sqrt{1 + \frac{1}{n} + \frac{(X_0 - \overline{X})^2}{S_{xx}}}$$

Example:

Continuing from the above illustration, we have $\nu = 6 - 2 = 4$ and we choose $\alpha = 0.05$. From Table B11 (p. 183), $t = 2.776$.
For the β_1 confidence interval, use Equation (10):

$$12.29 \pm 2.776 \, (42)/\sqrt{4230.8}$$

$$12.29 \pm 1.79$$

$$\therefore \qquad 10.50 < \beta_1 < 14.08$$

Next, at $X_0 = 0$, we compute the confidence interval on β_0 using Equation (11):

$$10.17 + 12.90(0) \pm 2.776(42) \sqrt{\frac{1}{6} + \frac{(0 - 40.8)^2}{4230.8}}$$

$$10.77 \pm 87.26$$

$$\therefore \qquad -77.1 < \beta_0 < 97.43$$

Finally, for the confidence interval on Y at X_0, use Equation (12):

$$10.17 + 12.90(0) \pm 2.776(42) \sqrt{1 + \frac{1}{6} + \frac{(0 - 40.8)^2}{4230.8}}$$

$$10.17 \pm 145.6$$

$$\therefore \quad -135.4 < Y < 155.8$$

In the above example, the calculations were illustrated for $X_0 = 0$. We can get a better feel for the confidence interval over the entire range of the correlation by choosing X_0 differently. This shows the predictive Y confidence limit graphically over the entire range of the regression formula.

The method of least squares can also be applied to non-linear equations which are reducible to linear form. Refer to power law and exponential fits.

Linear regression lends itself to easy adaptation to personal computer spreadsheet programs such as Lotus 1-2-3. Although some effort is required initially to adapt a spreadsheet to perform regressions, once the calculations are performed for a sample case, the spreadsheet becomes a template that is ready for future use. The following discussions illustrate how such a template can be constructed on the Lotus 1-2-3 version 1A for three cases (linear, power-law and exponential fit regressions for one independent variable). All three can be designed onto a master template that can be generalized to perform the regressions simultaneously to determine the best fitting model for a given database. To construct the template we shall use some data obtained from studying the compression characteristics of catalyst granules generated in a fluid bed granulation process. In the data set, the percent friability of one type of catalyst (Y-dependent variable) is studied as a function of the compaction pressure (X-independent variable, MN/m^2).

- Start with a clean worksheet. Use the Worksheet-Erase-Yes command sequence (/WEY = Slash − Worksheet − Erase − Yes).
- Type the title "X" in cell A3 [refer to Table B5 (p. 176)]. The first two rows of the spreadsheet are left blank for labeling purposes at the end of the exercise.
- In a similar fashion, type "Y" in the B3 cell.
- Allow a space between the data titles and the tabulated data. Beginning with cells A5 and B5, type in the data. When inputing large bodies of data it is advantageous to repeatedly protect what has been typed by saving it to the data disk before completing the task. In this example we will

use a data set of seven, however, the data entry is simply a function of the disk capacity.

- Next, move the cursor to cell H5 and type "=n". Type in the series of statements in cells H6–H28 as shown in Table B5. These are comment statements which identify calculations that are to be performed in cells G5 through G28. As you enter information onto the spreadsheet, you will notice that the calculations require sums of X^2, Y^2, and the product XY. Use columns *M, N,* and *O* to perform these computations. In cell M3, enter the identifying label "X ^ 2"; in cell N3 type "Y ^ 2", in cell O3 type "X*Y".

  Use the Copy command (/C) to fill all the calculations for each column for as large a data set as you may envision using (e.g., 200 data points). The Copy command updates calculations down each column, for each new cell drawing upon the data values in columns A and B.

- Before performing the regression calculations, construct a scatter plot using the Graph command. Select the *XY* option after typing /GT (i.e., the Graph Type command). Choose the *x*-axis for the first variable with the /GX command. /GB command places the second variable on the *y*-axis. Command /GOF prevents 1-2-3 from tracing lines between data points. The program will display the menu Graph A B C D E F. Select one of these graphs pending the plotting symbol of preference, and strike the Enter key. A second level menu appears offering Line, Symbol, Both or Neither. Select Symbol and strike the Enter key. Striking /GV will enable us to view the scatter plot of column B versus column A data values.

  Title the plot to identify it for future references (e.g., Linear Regression Plot) with the /GOTF command sequence. Using /GOTS one can add a second title (e.g., the number of data sets in the analysis). Titles for the *X*- and *Y*-axis can also be introduced.

- Returning to the calculation rows in column G (Table B5), the following entries should be made using the Lotus calculating function symbol command @:

Cell	Function
G5	@ COUNT (A5..A200)
G6	@ SUM (A5..A200)
G7	@ SUM (B5..B200)
G8	@ SUM (M5..M200)
G9	@ SUM (N5..N200)
G10	@ SUM (O5..O200)

(continued)

Cell	Function
G11	+G6/G5
G12	+G7/G5
G13	+G8−G6^2/G5
G14	+G9−G7^2/G5
G15	+G10−G6*G7/G5
G17	+G15/G13
G18	+G12−G17*G11
G19	G15/@ SQRT (G13*Go14))
G22	@ SQRT ((1/(G5−2))*(G14−(G15^2/G13)))
G25	@ MIN (A5..A200)
G26	@ MAX (A5..A200)
G27	@ MIN (B5..B200)
G28	@ MAX (B5..B200)

- The above sequence of calculations produces the coefficients for the regression giving us the linear expression: $y = G18 + G17*X$. Use this expression to predict any y-value (e.g., % friability) for each X (compaction pressure). To compare predictions with measured values:
  - Enter PRED Y as a label in cell C3. In cell C5 enter: +G$17*A5+G$18. Using the Copy command (/C) repeat the calculation throughout the column.
  - Update the graphics with the /GA command by typing the range C5..C200. This command plots the predicted Y values onto the graph. To distinguish the predicted values from measured data, use the /GOFA command and move the cursor to the Line option in the menu and strike the Enter key. In this manner, the predicted y-values will appear as a continuous line and actual data as symbols scattered about the line fit.
- The correlation coefficient $r*$ is contained in cell G19. This coefficient provides a measure of the strength of the correlation.
- Confidence limits control the certainty with which the observed value will fall within a given range of the predicted mean that is generated in column C. The size of the range is expressed as standard deviations. Compute the standard deviations of the points around the regression line as follows. We start by defining cell G22:

$$G22 \text{ is } @SQRT [(1/(G5−2))* (G14−(G15^2/G13)))]$$

The size of this range determines the probability that an observed value will fall within it. Although the correlation coefficient in the example is high ($r* = 0.82$), at the extremes of the data the correlation may in fact not provide a good representation. For this reason, it is important to quantify the confidence intervals of the regression model. This can be done by applying a Student t statistic. In cells H3 and H4 (refer to Table B5) type the labels "=DEGREES OF FREEDOM" and "=STUDENT'S t", respectively. The corresponding value for the degrees of freedom is typed into cell G3 as: $+G5 - 2$ (i.e., the sample population is 2).

The value of the Student t statistic can be obtained from an appropriate statistics handbook or Table B11. For the example, obtain $t = 2.571$ for a 95% confidence level and 5 degrees of freedom. Hence, the confidence limit requires us to specify an appropriate t-value based on the size of the data set being evaluated.

With cells G4 and G22 defined, go to cell D3 and type label 2.5 to denote the 2.5th percentile. In cell D5 type:

$$+C5 - \$G\$4*(@SQRT(\$G\$22 \char`\^ 2*(1 + 1/\$G\$5$$

$$+(A5 - \$G\$11) \char`\^ 2/\$G\$12)))$$

The dollar sign ($) indicates the absolute reference calculated from the given data set. Copy D5 into D6..D200 to return to the lower percentile.

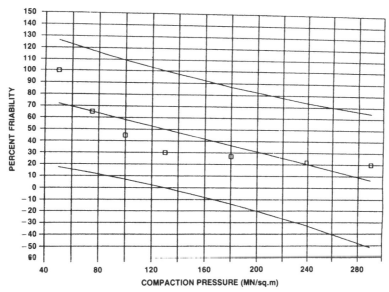

Figure B8. Regression example with confidence limits.

- Repeat the last step and define cell E3 as 97.5. In cell E5 type:

$$+C5 + \$G\$4*(@SQRT(\$G\$22 \hat{\ } 2*(1 + 1/\$G\$5$$

$$+(A5 - \$G\$11) \hat{\ } 2/\$G\$12)))$$

Copy the contents of E5 into E6..E200.
- Update the graph with columns D and E. The final product is a scatter plot showing the best line fit along with the limits for the 2.5 and 97.5 percentile, as shown in Figure B8.

Upon completion, we have created a template that can be reused with any data. By retyping new values for X and Y in columns A and B, the regression is automatically performed along with a plot. Similar routines can be prepared for power-law, exponential and quadratic regressions.

LOGARITHMIC CURVE FIT

Data following this relationship have the form:

$$y = a + b \ln x$$

to a set of data points

$$\{(x_i, y_i), i = 1,2, \ldots, n\}$$

The equation can be developed by linear regression where the output statistics are:

1. Regression coefficients:

$$b = \frac{\Sigma y_i \ln x_i - \dfrac{1}{n} \Sigma \ln x_i \Sigma y_i}{\Sigma (\ln x_i)^2 - \dfrac{1}{n} (\Sigma \ln x_i)^2}$$

$$a = \frac{1}{n} (\Sigma y_i - b \Sigma \ln x_i)$$

2. Coefficient of determination:

$$r^2 = \frac{\left[\Sigma \, y_i \ln x_i \; - \; \dfrac{1}{n} \, \Sigma \ln x_i \, \Sigma \, y_i \right]^2}{\left[\Sigma \, (\ln x_i)^2 \; - \; \dfrac{1}{n} \, (\Sigma \ln x_i)^2 \right] \left[\Sigma y_i^2 \; - \dfrac{1}{n} (\Sigma \, y_i)^2 \right]}$$

3. Estimated value of y (i.e., $\hat{y}$) for a given x:

$$\hat{y} = a + b \ln x$$

Note that n is a positive integer and $n \neq 1$.

LOGARITHMIC NORMAL DISTRIBUTION[21]

If x is a random variable whose logarithm is normally distributed with mean m and variance σ^2, then x has a logarithmic normal distribution with density function

$$f(x) = \frac{1}{x\sqrt{2\pi\sigma^2}} \, e^{-(1/2\sigma^2)(\ln x - m)^2}$$

where $x > 0$.

This method computes $f(x)$ and the following statistics for given m, σ^2:

$$\text{median} = e^m$$
$$\text{mode} = e^{m - \sigma^2}$$
$$\text{mean} = e^{m + (\sigma^2/2)}$$
$$\text{variance} = e^{\sigma^2 + 2m} \, (e^{\sigma^2} - 1)$$

MANN-WHITNEY STATISTIC[22]

The Mann-Whitney test statistic is designed for testing the null hypothesis of no difference between two populations. As outlined below, the test statistic

[21]Suggested reference—K. A. Brownlee, *Statistical Theory and Methodology in Science and Engineering*, John Wiley and Sons (1965).

[22]Suggested reference—J. E. Freund and R. E. Walpole, *Mathematical Statistics*, Prentice-Hall (1962). D. B. Owen, *Handbook of Statistical Tables*, Addison-Wesley (1962).

is performed on two independent samples of equal or unequal size. The Mann-Whitney test statistic is defined as:

$$U = n_1 \, n_2 + \frac{n_1 \, (n_1 + 1)}{2} - \sum_{i=1}^{n_i} R_i$$

where n_1 and n_2 are the sizes of the two samples. Arrange all values from both samples jointly (as if they were one sample) in an increasing order of magnitude; let R_i ($i = 1, 2, \ldots, n_1$) be the ranks assigned to the values of the first sample (it is immaterial which sample is referred to as the "first").

When n_1 and n_2 are small, the Mann-Whitney test bases on the exact distribution of U and specially constructed tables. When n_1 and n_2 are both large (i.e., >8) then:

$$z = \frac{U - \dfrac{n_1 \, n_2}{2}}{\sqrt{n_1 \, n_2 \, (n_1 + n_2 + 1)/12}}$$

is approximately a random variable having the standard normal distribution.

Example:

Given the following data sets:

Sample 1	14.9	11.3	13.2	16.6	17	14.1	15.4	13	16.9	
Rank R_i	7	1	4	12	14	5	10	10	13	
Sample 2	15.2	19.8	14.7	18.3	16.2	21.1	18.9	12.2	15.3	19.4
Rank	8	18	6	15	11	19	16	2	9	17

Solution:

$$n_1 = 9, \, n_2 = 10, \, U = 66.00, \, z = 1.71$$

MEAN-SQUARE SUCCESSIVE DIFFERENCE[23]

When test and estimation techniques are used, the method of drawing the sample from the population is specified to be random in most cases. If obser-

[23]Suggested reference—Dixon and Massey, *Introduction to Statistical Analysis*, McGraw-Hill (1969).

vations are chosen in a sequence $x_1, x_2, \ldots, x_n$, the mean-square successive difference can be used to test for randomness:

$$\eta = \sum_{i=1}^{n-1} (x_i - x_{i+1})^2 \Big/ \sum_{i=1}^{n} (x_i - \bar{x})^2$$

If n is large (i.e., >20) and the population is normal, then:

$$z = \frac{1 - \eta/2}{\sqrt{\dfrac{n - 2}{n^2 - 1}}}$$

has approximately the standard normal distribution. Long trends are associated with large positive values of z and short oscillations with large negative values.

Example:

Given the following data set

0.53, 0.52, 0.39, 0.49, 0.97, 0.29, 0.65, 0.30, 0.40,
0.06, 0.14, 0.16, 0.68, 0.22, 0.68, 0.08, 0.52, 0.50,
0.63, 0.20, 0.67, 0.44, 0.64, 0.40, 0.97, 0.03, 0.73,
0.24, 0.57, 0.35

compute the mean-square successive differences:

Solution:

$$n = 30, \eta = 2\text{-}81, z = 2.29$$

MEAN, STANDARD DEVIATION, STANDARD ERROR FOR GROUPED DATA

Given a set of data points

$$x_1, x_2, \ldots, x_n$$

with respective frequencies

$$f_1, f_2, \ldots, f_n$$

The following statistics are computed:

$$\text{mean } \bar{x} = \frac{\Sigma f_i x_i}{\Sigma f_i}$$

$$\text{standard deviation } s = \sqrt{\frac{\Sigma f_i x_i^2 - (\Sigma f_i)\bar{x}^2}{\Sigma f_i - 1}}$$

$$\text{standard error } s_{\bar{x}} = \frac{s_x}{\sqrt{\Sigma f_i}}$$

Example:

Given the data set:

x_i	2	3.4	7	11	23	3.41
f_i	5	3	4	2	3	1

Solution:

$\bar{x} = 7.92$, $s = 7.52$, $s_{\bar{x}} = 1.77$

MOMENTS, SKEWNESS AND KURTOSIS[24]

The following statistics for a set of given data $\{x_1, x_2, \ldots, s_n\}$ are applied:

1st moment

$$\bar{x} = \frac{1}{n} \sum_{i=1}^{n} x_i$$

2nd moment

$$m_2 = \frac{1}{n} \Sigma x_i^2 - \bar{x}^2$$

---

[24]Suggested reference—M. R. Spiegel, *Theory and Problems of Statistics*, Schaum's Outline, McGraw-Hill (1961).

3rd moment

$$m_3 = \frac{1}{n} \Sigma x_i^3 - \frac{3}{n} \bar{x} \Sigma x_i^2 + 2\bar{x}^3$$

4th moment

$$m_4 = \frac{1}{n} \Sigma x_i^4 - \frac{4}{n} \bar{x} \Sigma x_i^3 + \frac{6}{n} \bar{x}^2 \Sigma x_i^2 - 3\bar{x}^4$$

moment coefficient of skewness

$$\gamma_1 = \frac{m_3}{(m_2)^{3/2}}$$

moment coefficient of kurtosis

$$\gamma_2 = \frac{m_4}{(m_2)^2}$$

NESTED DESIGNS

Nested or hierarchical designs are based on a definite relationship between two factors. In this relationship, every level of Factor B appears with only one level of Factor A. This situation is characterized then with Factor B being nested in Factor A. As an example, consider the following response and factors:

RESPONSE – Green Strength of Compound Polymer
FACTOR A – Carbon Black (e.g., several suppliers used)
FACTOR B – Oil (several suppliers used)

The layout of the experimental design might take the form:

	Carbon Black		
	I	II	III
	1	4	7
Oils	2	5	8
	3	6	9

The factors can be described as being *crossed,* since all levels of Factor B appear with Factor A in a full factorial design. We can estimate the components of variance, where the overall relationship is:

$$\sigma^2 = \sigma_A^2 + \sigma_B^2 + \sigma_C^2 + \ldots$$

In the above illustration, we are interested in understanding why there is a variation in the compound green strength. The components of variance would therefore be assigned such that σ_A^2 is the carbon black variability, σ_B^2 is the oil variability (e.g., oil viscosity or purity level), and σ_C^2 is the test variability.

This type of analysis is sometimes referred to as a hierarchical design. The reason for this terminology is that Factor B is nested in Factor A, and in turn Factor C is in B, and so forth. The design establishes the number of nested levels of each factor n_A, n_B, n_C. The chart below shows the scheme for a three-factor nested design for illustration. As a general rule, to normalize the degrees of freedom for each factor, the number of levels for all but Factor A should be kept at two.

For a two-factor nested design (i.e., two components of variance), Factor B levels are essentially replicates of Factor A. Another way of viewing this is that Factor B is a random variable. The response model is:

$$Y_{ij} = \mu + A_i + B_{j(i)}$$

where

Y_{ij} = response at the ith level of Factor A and the jth nested condition of Factor B

μ = overall response mean

A_i = random variable with mean zero and variance σ_A^2 (where $i = 1, \ldots, n_A$)

$B_{u(i)}$ = random variable with mean zero and variance σ_B^2 (where $j = 1, \ldots, n_B$)

The response variation is thus:

$$\sigma_y^2 = \sigma_A^2 + \sigma_B^2$$

The standard deviations squared S_A^2 and S_B^2 must be estimated for σ_A^2 and σ_B^2, respectively.

The average response $\overline{Y}$ can be estimated from n_A levels of Factor A with n_B

levels of Factor B nested in each level of Factor A. The variance of $\overline{Y}$ can be computed from:

$$S_{\bar{y}}^2 = \frac{S_A^2}{n_A} + \frac{S_B^2}{n_A n_B}$$

The ANOVA scheme described earlier can be used to evaluate this type of problem.

For three components of variance (i.e., analysis of three-factor nested design), Factor C levels are nested in levels of Factor B, and Factor B levels are nested in levels of Factor A. That is, all three factors are essentially random variables in the components of the variance model. The response model is:

$$Y_{ijk} = \mu + A_i + B_{j(i)} + C_{k(ij)}$$

where

$Y_{ij}k$ = response at the ith level of A, jth level of B and at the kth level of C

μ = overall response mean

$A_i, B_{j(i)}, C_{k(ij)}$ = random variables with mean zero and variances $\sigma_A^2, \sigma_B^2, \sigma_C^2$, respectively

The subscripts i, j and k cover the run ranges $1 \ldots n_A$, $1 \ldots n_B$, and $1 \ldots n_C$, respectively.

For average response $\overline{Y}$ estimated from n_A levels of A, n_B levels of B, and n_C levels of C, the variance is:

$$S_{\bar{y}}^2 = \frac{S_A^2}{n_A} \frac{S_B^2}{n_A n_B} + \frac{S_C^2}{n_A n_B n_C}$$

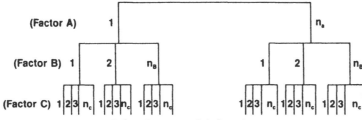

A three-factor nested design scheme.

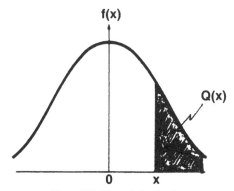

Figure B9. Normal distribution.

NORMAL DISTRIBUTION[25]
(see Figure B9)

The density function for a standard normal variable is:

$$f(x) = \frac{1}{\sqrt{2\pi}} \, e^{-(x^2/2)}$$

The upper tail area is:

$$Q(x) = \frac{1}{\sqrt{2\pi}} \int_x^\infty e^{-(t^2/2)} \, dt$$

For $x \geq 0$, polynomial approximation

$$Q(x) = f(x)(b_1 t + b_2 t^2 + b_3 t^3 + b_4 t^4 + b_5 t^5) + \epsilon(x)$$

where

$$|\epsilon(x)| < 7.5 \times 10^{-8}$$

$$t = \frac{1}{1 + rx}$$

[25]Suggested reference—Abramowitz and Stegun, *Handbook of Mathematical Functions*, National Bureau of Standards (1968).

$$r = 0.2316419$$

$$b_1 = .31938153$$

$$b_2 = -.356563782$$

$$b_3 = 1.781477937$$

$$b_4 = -1.821255978$$

$$b_5 = 1.330274429$$

ONE SAMPLE TEST STATISTICS FOR THE MEAN

For a normal population $(x_1, x_2 \ldots, x_n)$ with a known variance σ^2, a test of the null hypothesis

$$H_0: \text{mean } \mu = \mu_0$$

is based on the z statistic (which has a standard normal distribution)

$$z = \frac{\sqrt{n} \, (\bar{x} - \mu_0)}{\sigma}$$

If the variance σ^2 is unknown, then

$$t = \frac{\sqrt{n} \, (\bar{x} - \mu_0)}{s}$$

is used instead. This t statistic has the t distribution with $n - 1$ degrees of freedom. $\bar{x}$ and s are the sample mean and standard deviation, respectively.

Example:

Given the following data set

2.73, 0.45, 2.52, 1.19, 3.51, 2.75, 1.79, 1.83,
1, 0.87, 1.9, 1.62, 1.74, 1.92, 1.24, 2.68

Solution:

The test statistic is $t = -0.69$ and $z = -0.57$ if $T = 1$

PAIRED t STATISTIC[26]

Given a set of paired observations from two normal populations with means μ_1, μ_2 (unknown)

x_i	x_1	x_2	$\ldots$	x_n
y_i	y_1	y_2	$\ldots$	y_n

let

$$D_i = x_i - y_i$$

$$\overline{D} = \frac{1}{n} \sum_{i=1}^{n} D_i$$

$$s_D = \sqrt{\frac{\Sigma D_i^2 - \frac{1}{n}(\Sigma D_i)^2}{n-1}}$$

$$s_{\overline{D}} = \frac{s_D}{\sqrt{n}}$$

The test statistic

$$t = \frac{\overline{D}}{s_{\overline{D}}}$$

which has $(n - 1)$ degrees of freedom (df), can be used to test the null hypothesis

$$H_0: \mu_1 = \mu_2$$

[26]Suggested reference—B. Ostle, *Statistics in Research,* Iowa State University Press (1963).

Example:

$$x_i = 14 \quad 17.5 \quad 17 \quad 17.5 \quad 15.4$$
$$y_i = 17 \quad 20.7 \quad 21.6 \quad 20.9 \quad 17.2$$

Solution:

$$t = 7.16, \, df = 4.00$$

PARTIAL CORRELATION COEFFICIENTS[27]

The partial correlation coefficient measures the relationship between any two of the variables when all others are kept constant.

For the case of 3 variables, the partial correlation coefficient between X_1 and X_2 keeping X_3 constant is

$$r_{12 \cdot 3} = \frac{r_{12} - r_{13} r_{23}}{\sqrt{(1 - r_{13}^2)(1 - r_{23}^2)}}$$

where r_{ij} denotes the correlation coefficient of X_i and X_j.

Similarly, for the case of 4 variables, the partial correlation coefficient between X_1 and X_2 keeping X_3 and X_4 constant is

$$r_{12 \cdot 34} = \frac{r_{12 \cdot 4} - r_{13 \cdot 4} r_{23 \cdot 4}}{\sqrt{(1 - r_{13 \cdot 4}^2)(1 - r_{23 \cdot 4}^2)}} = \frac{r_{12 \cdot 3} - r_{14 \cdot 3} r_{24 \cdot 3}}{\sqrt{(1 - r_{14 \cdot 3}^2)(1 - r_{24 \cdot 3}^2)}}$$

Any partial correlation coefficient can be computed by means of these formulas (using this program) if correlation coefficients $r_{12}, r_{13}, r_{23}, \ldots$ are given.

PERMUTATION

A permutation is an ordered subset of a set of distinct objects. The number of possible permutations, each containing n objects, that can be formed from a collection of m distinct objects is given by:

$$_mP_n = \frac{m!}{(m - n)!} = m(m - 1) \ldots (m - n + 1)$$

[27]Suggested reference—S. Wilks, *Mathematical Statistics*, John Wiley and Sons (1962).

where m, n are integers and $0 \leq n \leq m$.
   Note that:

1. $_mP_n$ can also be denoted by P_n^m, $P(m,n)$ or $(m)_n$
2. $_mP_0 = 1$, $_mP_1 = m$, $_mP_m = m!$

POISSON DISTRIBUTION

The density function is:

$$f(x) = \frac{\lambda^x \, e^{-\lambda}}{x!}$$

where $\lambda > 0$ and $x = 0, 1, 2, \ldots$
   The cumulative distribution is:

$$P(x) = \sum_{k=0}^{n} f(k)$$

The objective is to evaluate $f(x)$ and $P(x)$ for a given λ using the recursive relation:

$$f(x + 1) = \frac{\lambda}{x + 1} f(x)$$

Note that:

1. $f(0) = P(0)$
2. mean = variance = X

POWER CURVE FIT[28]

Data following this relationship have the form:

$$y = ax^b \quad (a > 0)$$

[28]Suggested references—(a) K. A. Brownlee, *Statistical Theory and Methodology in Science and Engineering*, John Wiley and Sons, New York (1965). (b) N. P. Cheremisinoff, *Practical Statistics for Engineers and Scientists*, Technomic Publishing Co., Lancaster, PA (1987). (c) N. P. Cheremisinoff, "Spotlight on Environmental Software," *Pollution Engineering Magazine* (January 1987).

to a set of data points

$$\{(x_i, y_i), i = 1, 2, \ldots, n\}$$

where $x_i > 0$, $y_i > 0$.

By writing this equation as

$$\ln y = b \ln x + \ln a$$

the problem can be solved as a linear regression problem.

Output statistics are:

1. Regression coefficients

$$b = \frac{\Sigma\,(\ln x_i)(\ln y_i) - \dfrac{(\Sigma\,\ln x_i)(\Sigma\,\ln y_i)}{n}}{\Sigma\,(\ln x_i)^2 - \dfrac{(\Sigma\,\ln x_i)^2}{n}}$$

$$a = \exp\left[\frac{\Sigma\,\ln y_i}{n} - b\,\frac{\Sigma\,\ln x_i}{n}\right]$$

2. Coefficient of determination

$$r^2 = \frac{\left[\Sigma\,(\ln x_i)(\ln y_i) - \dfrac{(\Sigma\,\ln x_i)(\Sigma\,\ln y_i)}{n}\right]^2}{\left[\Sigma\,(\ln x_i)^2 - \dfrac{(\Sigma\,\ln x_i)^2}{n}\right]\left[\Sigma\,(\ln y_i)^2 - \dfrac{(\Sigma\,\ln y_i)^2}{n}\right]}$$

3. Estimated value of y (i.e., $\hat{y}$) for given x

$$\hat{y} = ax^b$$

where n is a positive integer and $n \neq 1$.

In the Linear Regression discussion, a spreadsheet template was constructed using Lotus 1-2-3.

It would be useful to perform both linear and non linear regressions in a single step to assess the best possible regression model for a given data set. Since the Lotus spreadsheet is gigantic, there is plenty of room to expand on the template without having to store each routine under a separate file name.

A power-law model has the form:

$$Y = kx^n$$

where k and n are coefficients of the expression. As in the first example, the method of least squares can be applied by rewriting the power-law expression in linear form:

$$\ln Y = \ln k + n \ln x$$

Or, using the notation for a straight line:

$$Y' = b_1 + b_0 x'$$

where

$x' = \ln x$
$y' = \ln y$
$b_1 = \text{intercept} = \ln k$
$b_0 = \text{slope} = n$

Select a clear field on the same spreadsheet (columns R through AG provide sufficient room to construct the power-law regression section of the template).

Table B6 (p. 178) shows the layout of the sub-template which is constructed as follows.

Use the /C command to copy columns C3 through E3 into cells T3 through V3, since we will reuse these calculations with a slight modification.

Use the /C command to copy the statistical formulas and identifying labels in cells G5 through H22 into columns X3 through Y22.

Finally, copy the x^2, y^2 and xy calculations from columns M, N, O into columns AE, AF, AG.

Note that in using the copy command to the new columns, the calculation formulas for each cell have been automatically updated by Lotus. In the linear regression model the computations were performed using the input values in columns A and B (refer to Table B5). With the above copying steps we have essentially shifted the computation field to a new section of the spreadsheet, where the updated cells now obtain the input values for columns R and S.

Type identifying labels LN(x) and LN(y) into cells R3 and S3, respectively. These columns will be used to calculate the natural logs from the raw input data of columns A, B.

Note that since the regression involves taking logarithms, a data input of zero or a negative value would be meaningless. To avoid inputing data which the model cannot handle we make use of a logic operation known as the IF formula which has the format:

If (Question, yes, no)
  1   2   3

The first section is similar to a question. Lotus 1-2-3 examines the question and responds with a yes or no (true or false).

In cells R5 and S5 type:

@IF(A5 = 0,0,@LN(A5))
@IF(B5 = 0,0,@LN(B5))

And copy these into A6..A200 and B6..B200.

In cells R5 and S5, we are asking whether the values of A5 and B5 are zero. If the values are zero, the answer would be yes; if the values are non-zero (positive), the answer would be no. When the answer is yes Lotus 1-2-3 exhibits the value from section 2 of the IF statement (i.e., a value of zero is assigned in columns R and S). If the answer is no Lotus 1-2-3 unveils the value in section 3, which contains the natural logarithms of the input data of columns A and B.

The same approach must be used in columns T through V. If values are zero, we cannot take the log, and hence, we assign a value of zero and disregard the input in the calculations. As will become apparent at the end of the article, the IF statement serves as a useful vehicle for clearing the data set and enabling new data input of any size population.

Type in the following formulas to complete the regression:

Cell	Formula
T5	@IF(B5 = 0,0,@EXP($X$17*R5 + $X$18))
U5	@IF(B5 = 0,0,T5 − $G$4*$X$22*@SQRT(1 + 1/$X$5 + (R5 − $X$11) ^ 2/$X$13))
V5	@IF(C5 = 0,0,T5 + $G$4*$X$22*@SQRT(1 + 1/$X$5 + (R5 − $X$11) ^ 2/$X$13))

Copy the above cells into the columns (i.e., /C T5..V5 into T6..V200).

The regression sub-template for the power-law fit is now complete. Results are shown in Table B7 (p. 179).

We can now design an exponential sub-template. The exponential model has the form:

$$Y = k_0 e^{ax}$$

or in linear notation:

$$\ln Y = \ln k_0 + ax$$

The construction of this portion of the template is almost trivial since many of the statistics have already been calculated for the first two regressions. The user can use the /C command to copy those columns needed for regression and edit the formulas appropriately. The formula edits required are as follows:

Cell	Formula	Function Performed
AJ5:	@IF(B5 = 0,0,@EXP($AN$17*A5 + $AN$18))	PRED Y
AK5:	@IF(B5 = 0,0,AJ5 − $G$4* (@SQRT($AN$22 ^ 2*(1 + 1/$G$5 + (A5 − $G$11) ^ 2/$G$13))))	2.5 Percentile
AL5:	@IF(B5 = 0,0,AJ5 + $G$4* (@SQRT($AN$22 ^ 2*(1 + 1/$G$5 + (A5 − $G$11) ^ 2/$G$13))))	97.5 Percentile
AM5:	+B5 − AJ5	Deviation
AT5:	+A5*S5	$x \ln y$
AN10:	@SUM(AT5..AT200)	$\Sigma\, x \ln y$
AN12:	+X7/G5	$\ln y$ mean
AN15:	+AN10 − G6*X7/G5	$\Sigma\, x \ln y - \Sigma x\, \Sigma \ln y/n$
AN17:	+AN15/AN13	slope
AN18:	+AN12 − AN17*G11	intercept
AN19:	+AN15/@SQRT(G13*X14))	correlation coefficient
AN22:	@EXP(@SQRT((1/(G5 − 2))* (X14 − (AN15 ^ 2/G13))))	standard deviation

Table B8 (p. 181) shows the spreadsheet layout and results for the exponential regression.

As a final function for the template, we can plot all three regressions along with the raw data to obtain a visual assessment of the best fit from these models.

- Using /G—use the Reset (R) option to return the graph to its default mode.
- Hit TX (for Type Graph and XY plot option).
- Plot the values:

  x-axis A5..A200
  A.range—B5..B200 (raw *Y* data)
  B.range—C5..C200 (linear model predictions)
  C.range—T5..T200 (power-law prediction)
  D.range—AJ5..AJ200 (exponential)

- Use the OF (Options-Format) command and specify:

  'AS'—a data range—symbols.

  and for ranges B through D use both lines and symbols.
- Finally, use the legend command L to identify what each symbol refers to on the graph.

The end result is a plot showing all three regression fits along with the data as shown in Figure B10. For the example in this article we can see that the power-law model provides the best fit of the data.

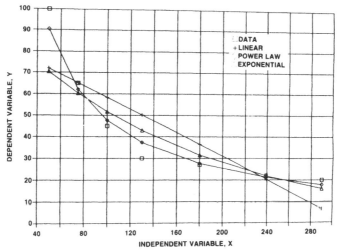

Figure B10. Regression examples.

Save the graph and one more step is needed to make the template universal.

A problem with the spreadsheet template in its present form is that the next time we go to use it, the population size must be greater than or equal to 7 (the amount used in the example). Any extraneous data must be cleared from input columns A and B along with the Student-t values in cell G4.

The template can be easily automated by the use of a macro. A macro is a string of commands placed in a column of cells on the worksheet. Until now, we have typed in the commands from the keyboard.

Move the cursor to a free cell on the worksheet. If you've followed this example, cell AV3 is a likely location. Enter the following line:

$$'/REA5..B200 \sim /REG4 \sim$$

The label-prefix ' (apostrophe) precedes all lines in a macro. The command typed here is: Slash $(/)$ — Range (R) — Erase (E) — A5..B200 — Return $(\sim)$ — Slash $(/)$ — Range (R) — Erase (E) — G4 — Return $(\sim)$. Collectively, the line instructs 1-2-3 to erase the data in columns A and B as well as the Student-t value in cell G4. In addition, since we used IF statements in the power-law (columns R through V) and exponential (Columns AJ–AL) models, values used in the regressions are automatically assigned zeros since an empty cell (columns A and B) is interpreted as 0 by Lotus. Hence, the power-law and exponential templates are automatically clear for new data entries.

Once the macro has been built it must be given a name. The macro will be referenced by this name. The Range Name command is requisitioned for this step. Type /R, and from the Lotus 1-2-3 menu select the name command (type N). To create a range name, type C and 1-2-3 will request the specifications. For the name enter \ E. Lotus will acknowledge the name and ask for coordinates of the range being named which in this case is cell AV3 (simply hit the return key). Macro names consist of the \ (backslash) followed by a single letter. This macro was named \ E since E stands for erase and is easy to remember. Table B8 summarizes the steps needed in creating a macro.

Once the macro has been named, it may be triggered anywhere on the worksheet by holding down the <Alt> key and pressing the letter of the macro. Hence, to trigger the macro, type <Alt> — E, and Lotus will erase the data in columns A and B and cell G4.

We can now dress up the program with appropriate identifying labels to distinguish which model is which (use rows 1 and 2) and save the program for future use.

In summary, a generalized template has been created which will provide simultaneous regressions of up to 195 data values (more if you like) for linear, power-law and exponential models. The user can automatically clear the data field at the start of each analysis, preparing the template for new data input. All three model results can be viewed on the same worksheet along with a comparative plot of the regressions and raw data which are automatically updated by Lotus for each new data base.

Using the principles outlined in this article and in the author's book [Reference 29(b)], the reader can expand the capabilities of the program to polynomial and multivariable non-linear regressions.

POLYNOMIAL CURVE FIT

A quadratic fit can be done on Lotus 1-2-3 by preparing a template. Refer to the Linear Regression example to see how to construct a regression routine template. If you have version 2 of Lotus, the program already has the capability for linear regression, which can be retrofitted to handle a polynomial. The following example illustrates how to construct a four term polynomial template.

Example:

The following temperatures were recorded at the die of an extruder for different values of the screw speed in rpm:

Screw speed (rpm), X	0	31	76	100	159	202
Die temperature (°C), Y	89	89	90	95	104	113

Regress the data to a polynomial fit (i.e., $Y = aX^n + bx^{n-1} + \ldots + cX + d$)

Solution:

The construction of the template on the Lotus spreadsheet is shown below. Type the titles shown in cells A1 through G1 to identify data and calculations. Columns A and B are for data (i.e., input screw speed data in cells A2 . . . A7 and die temperature data in cells B2 . . . B7). In this example, a four term polynomial is constructed. Type the following math executions:

> in Cell C2, type " ⌐ A2↑4"
> in Cell D2, type "+A2↑3"
> in Cell E2, type "+A2↑2"
> in Cell F2, type "+1*A2"

Use the copy command to fill in the calculations for all cells—type /C C2 . . . F2 . . . F7 ~ (note: " ~ " means return).

You are now ready to use the Lotus regression subroutine. Type /DR (slash, data, regress). Following the Lotus submenu, type "O" (for output) and move the cursor to cell H2 and hit return. Specify the X-input as cells C2 . . . F7 and the Y-input as B2 . . . B7. For the Intercept calculation, specify "C" for compute. Type "G" for go and the regression is performed by Lotus, reporting the results in cells H2 through M10. The dependent variable coefficients are reported in cells J9 through M9, and the intercept value is in cell K3. The coefficient of fit is given in cell K5.

In cell G1, type the title Ycalc and compute the correlation's Y values by typing in cell G2:

$$+\$J9*C2+\$K\$9*D2+\$L\$9*E2+\$M\$9*F2+\$K\$3$$

Use the /C command to repeat the calculations in cell G3 . . . G7. You can now construct a graph to compare the correlation fit to the data as shown in Figure B11. Refer to the Linear Regression example for details on constructing a graph on Lotus. By saving the worksheet, a template is created for reuse with new data. The template can be modified to provide calculations auto-

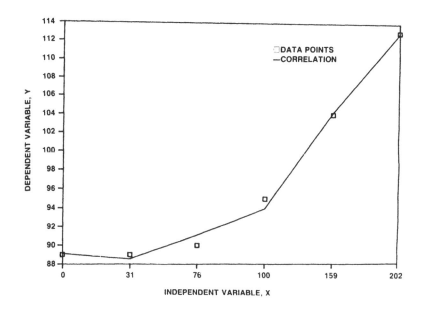

Figure B11. Example of polynomial fit on LOTUS.

matically by constructing a Macro. Refer to Linear and Power-Law Regression examples for a description on constructing macros.

The final equation in this example is:

$$Y = -4.83 \times 10^{-9}X^4 - 2.01 \times 10^{-7}X^3 + 1.08 \times 10^{-3}X^2$$

$$-5.42 \times 10^{-2}X + 89.13$$

PROBABILITY OF NO REPETITIONS IN A SAMPLE[29]

Suppose a sample of size n is drawn with replacement from a population containing m different objects. Let P be the probability that there are no repetitions in the sample, then:

$$P = \left(1 - \frac{1}{m}\right)\left(1 - \frac{2}{m}\right) \cdots \left(1 - \frac{n-1}{m}\right)$$

given integers m, n such that $m \geq n \geq 1$.

PROBIT ANALYSIS

This is a technique for assessing dose-response relationships. The response in this problem class is quantal (i.e., all-or-nothing). The probit model is based on the assumption that the tolerance distribution is Gaussian in either dose or log (dose).

The design is formulated such that several doses are chosen with the 50% response dose bracketed. If the 50% dose is not known, divide the N subjects equally among the selected doses. For large N, it's possible to run a pilot experiment with $N/10$ subjects to obtain an approximation of the 50% dosage. The interval of doses should be wide enough such that the extremes include the 5–15% and the 85–95% response ranges.

[29]Suggested reference—E. Parzen, *Modern Probability Theory and its Applications,* John Wiley and Sons (1960).

The analysis scheme is a weighted regression, where the highest weighting occurs for responses close to the 50% response and tailing off in either direction. The estimation procedure is iterative (described in standard statistics textbooks), where doses giving 0 or 100% response are ignored in the scheme.

SPEARMAN'S RANK CORRELATION COEFFICIENT[30]

Spearman's rank correlation coefficient is defined by:

$$r_s = 1 - \frac{6 \sum_{i=1}^{n} D_i^2}{n(n^2 - 1)}$$

where

n = number of paired observations (x_i, y_i)
D_i = rank (x_i) − rank (y_i) = $R_i - S_i$

If the X and Y random variables from which these n pairs of observations are derived are independent, then r_s has zero mean and a variance:

$$\frac{1}{n - 1}$$

A test for the null hypothesis:

$$H_0: X, \ Y \text{ are independent}$$

is using:

$$z = r_s \sqrt{n - 1}$$

which is approximately a standardized normal variable (for large n, say $n \geq 10$).

If the null hypothesis of independence is not rejected, we can infer that the population correlation coefficient $\varrho(x,y) = 0$, but dependence between the variables does not necessarily imply that $\varrho(x,y) \neq 0$.

[30]Suggested reference—J. D. Gibbons, *Nonparametric Statistical Inference*, McGraw-Hill (1971).

Note:

$$-1 \le r_s \le 1$$

where $r_s = 1$ indicates complete agreement in order of the ranks and $r_s = -1$ indicates complete agreement in the opposite order of the ranks.

STANDARD ERRORS FOR LINEAR REGRESSION[31]

Suppose $y = a_0 + a_1 x$ is the least squares fit to a set of data points $\{(x_i, y_i), i = 1, 2, \ldots, n\}$ and $\hat{y}$ is the estimated value on the line for a given x value. The principle formulae are:

1. Standard error of estimate (of y on x)

$$s_{y \cdot x} = \sqrt{\frac{\Sigma(y_i - \hat{y}_i)^2}{n - 2}}$$

$$= \sqrt{\frac{\Sigma y_i^2 - a_0 \Sigma y_i - a_1 \Sigma x_i y_i}{n - 2}}$$

2. Standard error of the regression coefficient a_0

$$s_0 = s_{y \cdot x} \sqrt{\frac{\Sigma x_i^2}{n \left[\Sigma x_i^2 - \frac{(\Sigma x_i)^2}{n} \right]}}$$

3. Standard error of the regression coefficient a_1

$$s_1 = \frac{s_{y \cdot x}}{\sqrt{\Sigma x_i^2 - \frac{(\Sigma x_i)^2}{n}}}$$

Note that n is a positive integer and $n \ne 1$ or 2.

[31]Suggested reference—Draper and Smith, *Applied Regression Analysis*, John Wiley and Sons (1966).

Example:

y_i:	92	85	78	81	54	51	40
x_i:	26	30	44	50	62	68	74

$$A_0 = 121.04$$
$$A_1 = -1.03$$

Solution:

$$s_{y \cdot x} = 6.34, \; s_0 = 7.47, \; s_1 = 0.14$$

STANDARDIZED SCORES

Given a set of data $\{x_1, x_2, \ldots, x_n\}$, this method finds $\{y_1, y_2, \ldots, y_n\}$ such that

$$y_i = \frac{x_i - \bar{x}}{s}$$

for $i = 1, 2, \ldots n$

where $\bar{x}$ and s are sample mean and standard deviation, respectively, of $\{x_1, x_2, \ldots, x_n\}$. $\{y_1, y_2, \ldots, y_n\}$ has mean zero and its standard deviation is 1.

This method can also transform $y_i's$ to $z_i's$ such that $\{z_1, z_2, \ldots, z_n\}$ has mean μ and standard deviation σ (μ and σ are given).

$$z_i = \sigma y_i + \mu$$

for $i = 1, 2, \ldots n$

t DISTRIBUTION[32]

The integral for t distribution is:

$$I(x, \nu) = \int_{-x}^{x} \frac{\Gamma\left(\frac{\nu + 1}{2}\right)\left(1 + \frac{y^2}{\nu}\right)^{-(\nu+1/2)}}{\sqrt{\pi \nu} \; \Gamma\left(\frac{\nu}{2}\right)} dy$$

[32]Suggested reference—Abramowitz and Stegun, *Handbook of Mathematical Functions*, National Bureau of Standards (1968).

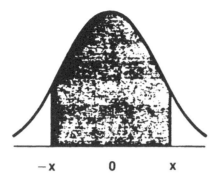

Figure B12. *t* distribution.

where $x > 0$, ν is the degrees of freedom (refer to Figure B12).
Formulas used are:

1. ν even

$$I(x,\nu) = \sin \theta \left\{ 1 + \frac{1}{2} \cos^2 \theta + \frac{1 \cdot 3}{2 \cdot 4} \cos^4 \theta + \ldots \right.$$

$$\left. + \frac{1 \cdot 3 \cdot 5 \ldots (\nu - 3)}{2 \cdot 4 \cdot 6 \ldots (\nu - 2)} \cos^{\nu-2} \theta \right\}$$

2. ν odd

$$I(x,\nu) = \begin{cases} \dfrac{2\theta}{\pi} & \text{if } \nu = 1 \\[2ex] \dfrac{2\theta}{\pi} + \dfrac{2}{\pi} \cos \theta \left\{ \sin \theta \left[1 + \dfrac{2}{3}\cos^2 \theta + \ldots \right.\right. \\[2ex] \left.\left. + \dfrac{2 \cdot 4 \ldots (\nu - 3)}{1 \cdot 3 \ldots (\nu - 2)} \cos^{\nu-3} \theta \right] \right\} & \text{if } \nu > 1 \end{cases}$$

where

$$\theta = \tan^{-1} \left(\frac{x}{\sqrt{\nu}} \right)$$

t STATISTIC FOR TWO MEANS[33]

Suppose $\{x_1, x_2, \ldots, x_{n_1}\}$ and $\{y_1, y_2, \ldots, y_{n_2}\}$ are independent random samples from two normal populations having means μ_1, μ_2 (unknown) and the same unknown variance σ^2.

We want to test the null hypothesis

$$H_0\!: \mu_1 - \mu_2 = D$$

where D is a given number.

Define

$$\bar{x} = \frac{1}{n_1} \sum_{i=1}^{n_1} x_i$$

$$\bar{y} = \frac{1}{n_2} \sum_{i=1}^{n_2} x_i$$

$$t = \frac{\bar{x} - \bar{y} - D}{\sqrt{\dfrac{1}{n_1} + \dfrac{1}{n_2}} \; \sqrt{\dfrac{\Sigma x_i^2 - n_1 \bar{x}^2 + \Sigma y_i^2 - n_2 \bar{y}^2}{n_1 + n_2 - 2}}}$$

We can use this t statistic, which has the t distribution with $n_1 + n_2 - 2$ degrees of freedom, to test the null hypothesis H_0.

Example:

$x = 79, 84, 108, 114, 120, 103, 122, 120$
$y = 91, 103, 90, 113, 108, 87, 100, 80, 99, 54$
$n_1 = 8$
$n_2 = 10$

Solution:

If $D = 0$ (i.e., $H_0\!: \mu_1 = \mu_2$), then $\bar{x} = 106.25$, $\bar{y} = 92.5$, $t = 1.73$

[33]Suggested reference—K. A. Brownlee, *Statistical Theory and Methodology in Science and Engineering*, John Wiley and Sons (1965).

TRIMMING

This is a technique aimed at eliminating the extreme values at each end of a sample. This method is applied when individuals in the sample are suspected of having an origin from a different population. In this case, we may want to estimate the mean in a manner to minimize the effects of these extraneous data. Another similar method is *Wintorization*. In this procedure, the smallest and largest observations are assigned the value of their closest neighbors. Both are illustrated by the sample data set:

$$X_{(1)} = 67.9 \quad X_{(3)} = 73.7 \quad X_{(5)} = 85.5$$

$$X_{(2)} = 68.2 \quad X_{(4)} = 78.6 \quad X_{(6)} = 85.9$$

$$\overline{X} = 459.8/6 = 76.63$$

For trimming:

$$\overline{X} = (68.2 + 73.7 + 78.6 + 85.5)/4 = 76.50$$

For Wintorization:

$$\overline{X} = (68.2 + 68.2 + 73.7 + 78.6 + 85.5 + 85.5)/6 = 76.62$$

WEIBULL DISTRIBUTION PARAMETER CALCULATION

The Weibull probability density function is given by:

$$f(x) = \frac{bx^{(b-1)}}{\theta^b} e^{-(x/\theta)^b}$$

where $\theta > 0, b > 0, x > 0$.

The cumulative distribution function is:

$$F(x) = 1 - e^{-(x/\theta)^b}$$

For a set of data $\{x_1, \ldots, x_n\}$, the Weibull parameters b and θ are to be calculated for these functions.

A common application is to use Weibull analysis for failure data where all samples are tested to failure.

The median rank (M.R.) is calculated by:

$$\frac{R_i - 0.3}{n + 0.4}$$

where R_i is the rank of failure data x_i. Using this median rank as an approximation of $F(x_i)$, a least squares fit is performed to the linearized form of the cumulative distribution function

$$\ln \ln \left(\frac{1}{1 - F(x)} \right) = b \ln x - b \ln \theta$$

The solution is similar to the linear regression problem, and estimates of b and θ are obtained.

APPENDIX B—STATISTICS CHARTS AND TABLES

Table B1. One-sided F distribution.

$\alpha = 0.10$

ν_D \ ν_N	1	2	3	4	5	6	7	8	9	10	12	15	20	24	30	40	60	120	∞
1	39.86	49.50	53.59	55.83	57.24	58.20	58.91	59.44	59.86	60.19	60.71	61.22	61.74	62.00	62.26	62.53	62.79	63.06	63.33
2	8.53	9.00	9.16	9.24	9.29	9.33	9.35	9.37	9.38	9.39	9.41	9.42	9.44	9.45	9.46	9.47	9.47	9.48	9.49
3	5.54	5.46	5.39	5.34	5.31	5.28	5.27	5.25	5.24	5.23	5.22	5.20	5.18	5.18	5.17	5.16	5.15	5.14	5.13
4	4.54	4.32	4.19	4.11	4.05	4.01	3.98	3.95	3.94	3.92	3.90	3.87	3.84	3.83	3.82	3.80	3.79	3.78	3.76
5	4.06	3.78	3.62	3.52	3.45	3.40	3.37	3.34	3.32	3.30	3.27	3.24	3.21	3.19	3.17	3.16	3.14	3.12	3.10
6	3.78	3.46	3.29	3.18	3.11	3.05	3.01	2.98	2.96	2.94	2.90	2.87	2.84	2.82	2.80	2.78	2.76	2.74	2.72
7	3.59	3.26	3.07	2.96	2.88	2.83	2.78	2.75	2.72	2.70	2.67	2.63	2.59	2.58	2.56	2.54	2.51	2.49	2.47
8	3.46	3.11	2.92	2.81	2.73	2.67	2.62	2.59	2.56	2.54	2.50	2.46	2.42	2.40	2.38	2.36	2.34	2.32	2.29
9	3.36	3.01	2.81	2.69	2.61	2.55	2.51	2.47	2.44	2.42	2.38	2.34	2.30	2.28	2.25	2.23	2.21	2.18	2.16
10	3.29	2.92	2.73	2.61	2.52	2.46	2.41	2.38	2.35	2.32	2.28	2.24	2.20	2.18	2.16	2.13	2.11	2.08	2.06
11	3.23	2.86	2.66	2.54	2.45	2.39	2.34	2.30	2.27	2.25	2.21	2.17	2.12	2.10	2.08	2.05	2.03	2.00	1.97
12	3.18	2.81	2.61	2.48	2.39	2.33	2.28	2.24	2.21	2.19	2.15	2.10	2.06	2.04	2.01	1.99	1.96	1.93	1.90
13	3.14	2.76	2.56	2.43	2.35	2.28	2.23	2.20	2.16	2.14	2.10	2.05	2.01	1.98	1.96	1.93	1.90	1.88	1.85
14	3.10	2.73	2.52	2.39	2.31	2.24	2.19	2.15	2.12	2.10	2.05	2.01	1.96	1.94	1.91	1.89	1.86	1.83	1.80
15	3.07	2.70	2.49	2.36	2.27	2.21	2.16	2.12	2.09	2.06	2.02	1.97	1.92	1.90	1.87	1.85	1.82	1.79	1.76
16	3.05	2.67	2.46	2.33	2.24	2.18	2.13	2.09	2.06	2.03	1.99	1.94	1.89	1.87	1.84	1.81	1.78	1.75	1.72
17	3.03	2.64	2.44	2.31	2.22	2.15	2.10	2.06	2.03	2.00	1.96	1.91	1.86	1.84	1.81	1.78	1.75	1.72	1.69
18	3.01	2.62	2.42	2.29	2.20	2.13	2.08	2.04	2.00	1.98	1.93	1.89	1.84	1.81	1.78	1.75	1.72	1.69	1.66
19	2.99	2.61	2.40	2.27	2.18	2.11	2.06	2.02	1.98	1.96	1.91	1.86	1.81	1.79	1.76	1.73	1.70	1.67	1.63
20	2.97	2.59	2.38	2.25	2.16	2.09	2.04	2.00	1.96	1.94	1.89	1.84	1.79	1.77	1.74	1.71	1.68	1.64	1.61
21	2.96	2.57	2.36	2.23	2.14	2.08	2.02	1.98	1.95	1.92	1.87	1.83	1.78	1.75	1.72	1.69	1.66	1.62	1.59
22	2.95	2.56	2.35	2.22	2.13	2.06	2.01	1.97	1.93	1.90	1.86	1.81	1.76	1.73	1.70	1.67	1.64	1.60	1.57
23	2.94	2.55	2.34	2.21	2.11	2.05	1.99	1.95	1.92	1.89	1.84	1.80	1.74	1.72	1.69	1.66	1.62	1.59	1.55
24	2.93	2.54	2.33	2.19	2.10	2.04	1.98	1.94	1.91	1.88	1.83	1.78	1.73	1.70	1.67	1.64	1.61	1.57	1.53
25	2.92	2.53	2.32	2.18	2.09	2.02	1.97	1.93	1.89	1.87	1.82	1.77	1.72	1.69	1.66	1.63	1.59	1.56	1.52
26	2.91	2.52	2.31	2.17	2.08	2.01	1.96	1.92	1.88	1.86	1.81	1.76	1.71	1.68	1.65	1.61	1.58	1.51	1.50
27	2.90	2.51	2.30	2.17	2.07	2.00	1.95	1.91	1.87	1.85	1.80	1.75	1.70	1.67	1.64	1.60	1.57	1.53	1.49
28	2.89	2.50	2.29	2.16	2.06	2.00	1.94	1.90	1.87	1.84	1.79	1.74	1.69	1.66	1.63	1.59	1.56	1.52	1.48
29	2.89	2.50	2.28	2.15	2.06	1.99	1.93	1.89	1.86	1.83	1.78	1.73	1.68	1.65	1.62	1.58	1.55	1.51	1.47
30	2.88	2.49	2.28	2.14	2.05	1.98	1.93	1.88	1.85	1.82	1.77	1.72	1.67	1.64	1.61	1.57	1.54	1.50	1.46
40	2.84	2.41	2.23	2.09	2.00	1.93	1.87	1.83	1.79	1.76	1.71	1.66	1.61	1.57	1.54	1.51	1.47	1.42	1.38
60	2.79	2.39	2.18	2.04	1.95	1.87	1.82	1.77	1.74	1.71	1.66	1.60	1.54	1.51	1.48	1.44	1.40	1.35	1.29
120	2.75	2.35	2.13	1.99	1.90	1.82	1.77	1.72	1.68	1.65	1.60	1.55	1.48	1.45	1.41	1.37	1.32	1.26	1.19
∞	2.71	2.30	2.08	1.94	1.85	1.77	1.72	1.67	1.63	1.60	1.55	1.49	1.42	1.38	1.34	1.30	1.24	1.17	1.00

166

Table B1. (continued).

α = 0.05

ν_D \ ν_N	1	2	3	4	5	6	7	8	9	10	12	15	20	24	30	40	60	120	∞
1	161.4	199.5	215.7	224.6	230.2	234.0	236.8	238.9	240.5	241.9	243.9	245.9	248.0	249.1	250.1	251.1	252.2	253.3	254.3
2	18.51	19.00	19.16	19.25	19.30	19.33	19.35	19.37	19.38	19.40	19.41	19.43	19.45	19.46	19.46	19.47	19.48	19.49	19.50
3	10.13	9.55	9.28	9.12	9.01	8.94	8.89	8.85	8.81	8.79	8.74	8.70	8.66	8.64	8.62	8.59	8.57	8.55	8.53
4	7.71	6.94	6.59	6.39	6.26	6.16	6.09	6.04	6.00	5.96	5.91	5.86	5.80	5.77	5.75	5.72	5.69	5.66	5.63
5	6.61	5.79	5.41	5.19	5.05	4.95	4.88	4.82	4.77	4.74	4.68	4.62	4.56	4.53	4.50	4.46	4.43	4.40	4.36
6	5.99	5.14	4.76	4.53	4.39	4.28	4.21	4.15	4.10	4.06	4.00	3.94	3.87	3.84	3.81	3.77	3.74	3.70	3.67
7	5.59	4.74	4.35	4.12	3.97	3.87	3.79	3.73	3.68	3.64	3.57	3.51	3.44	3.41	3.38	3.34	3.30	3.27	3.23
8	5.32	4.46	4.07	3.84	3.69	3.58	3.50	3.44	3.39	3.35	3.28	3.22	3.15	3.12	3.08	3.04	3.01	2.97	2.93
9	5.12	4.26	3.86	3.63	3.48	3.37	3.29	3.23	3.18	3.14	3.07	3.01	2.94	2.90	2.86	2.83	2.79	2.75	2.71
10	4.96	4.10	3.71	3.48	3.33	3.22	3.14	3.07	3.02	2.98	2.91	2.85	2.77	2.74	2.70	2.66	2.62	2.58	2.54
11	4.84	3.98	3.59	3.36	3.20	3.09	3.01	2.95	2.90	2.85	2.79	2.72	2.65	2.61	2.57	2.53	2.49	2.45	2.40
12	4.75	3.89	3.49	3.26	3.11	3.00	2.91	2.85	2.80	2.75	2.69	2.62	2.54	2.51	2.47	2.43	2.38	2.34	2.30
13	4.67	3.81	3.41	3.18	3.03	2.92	2.83	2.77	2.71	2.67	2.60	2.53	2.46	2.42	2.38	2.34	2.30	2.25	2.21
14	4.60	3.74	3.34	3.11	2.96	2.85	2.76	2.70	2.65	2.60	2.53	2.46	2.39	2.35	2.31	2.27	2.22	2.18	2.13
15	4.54	3.68	3.29	3.06	2.90	2.79	2.71	2.64	2.59	2.54	2.48	2.40	2.33	2.29	2.25	2.20	2.16	2.11	2.07
16	4.49	3.63	3.24	3.01	2.85	2.74	2.66	2.59	2.54	2.49	2.42	2.35	2.28	2.24	2.19	2.15	2.11	2.06	2.01
17	4.45	3.59	3.20	2.96	2.81	2.70	2.61	2.55	2.49	2.45	2.38	2.31	2.23	2.19	2.15	2.10	2.06	2.01	1.96
18	4.41	3.55	3.16	2.93	2.77	2.66	2.58	2.51	2.46	2.41	2.34	2.27	2.19	2.15	2.11	2.06	2.02	1.97	1.92
19	4.38	3.52	3.13	2.90	2.74	2.63	2.54	2.48	2.42	2.38	2.31	2.23	2.16	2.11	2.07	2.03	1.98	1.93	1.88
20	4.35	3.49	3.10	2.87	2.71	2.60	2.51	2.45	2.39	2.35	2.28	2.20	2.12	2.08	2.04	1.99	1.95	1.90	1.84
21	4.32	3.47	3.07	2.84	2.68	2.57	2.49	2.42	2.37	2.32	2.25	2.18	2.10	2.05	2.01	1.96	1.92	1.87	1.81
22	4.30	3.44	3.05	2.82	2.66	2.55	2.46	2.40	2.34	2.30	2.23	2.15	2.07	2.03	1.98	1.94	1.89	1.84	1.78
23	4.28	3.42	3.03	2.80	2.64	2.53	2.44	2.37	2.32	2.27	2.20	2.13	2.05	2.01	1.96	1.91	1.86	1.81	1.76
24	4.26	3.40	3.01	2.78	2.62	2.51	2.42	2.36	2.30	2.25	2.18	2.11	2.03	1.98	1.94	1.89	1.84	1.79	1.73
25	4.24	3.39	2.99	2.76	2.60	2.49	2.40	2.34	2.28	2.24	2.16	2.09	2.01	1.96	1.92	1.87	1.82	1.77	1.71
26	4.23	3.37	2.98	2.74	2.59	2.47	2.39	2.32	2.27	2.22	2.15	2.07	1.99	1.95	1.90	1.85	1.80	1.75	1.69
27	4.21	3.35	2.96	2.73	2.57	2.46	2.37	2.31	2.25	2.20	2.13	2.06	1.97	1.93	1.88	1.84	1.79	1.73	1.67
28	4.20	3.34	2.95	2.71	2.56	2.45	2.36	2.29	2.24	2.19	2.12	2.04	1.96	1.91	1.87	1.82	1.77	1.71	1.65
29	4.18	3.33	2.93	2.70	2.55	2.43	2.35	2.28	2.22	2.18	2.10	2.03	1.94	1.90	1.85	1.81	1.75	1.70	1.64
30	4.17	3.32	2.92	2.69	2.53	2.42	2.33	2.27	2.21	2.16	2.09	2.01	1.93	1.89	1.84	1.79	1.74	1.68	1.62
40	4.08	3.23	2.84	2.61	2.45	2.34	2.25	2.18	2.12	2.08	2.00	1.92	1.84	1.79	1.74	1.69	1.64	1.58	1.51
60	4.00	3.15	2.76	2.53	2.37	2.25	2.17	2.10	2.04	1.99	1.92	1.84	1.75	1.70	1.65	1.59	1.53	1.47	1.39
120	3.92	3.07	2.68	2.45	2.29	2.17	2.09	2.02	1.96	1.91	1.83	1.75	1.66	1.61	1.55	1.50	1.43	1.35	1.25
∞	3.84	3.00	2.60	2.37	2.21	2.10	2.01	1.94	1.88	1.83	1.75	1.67	1.57	1.52	1.46	1.39	1.32	1.22	1.00

(continued)

167

Table B1. *(continued).*

α = 0.01

ν_D \\ ν_N	1	2	3	4	5	6	7	8	9	10	12	15	20	24	30	40	60	120	∞
1	4052	4999.5	5403	5625	5761	5859	5928	5982	6022	6056	6106	6157	6209	6235	6261	6287	6313	6339	6366
2	98.50	99.00	99.17	99.25	99.30	99.33	99.36	99.37	99.39	99.40	99.42	99.43	99.45	99.46	99.47	99.47	99.48	99.49	99.50
3	34.12	30.82	29.46	28.71	28.24	27.91	27.67	27.49	27.35	27.23	27.05	26.87	26.69	26.60	26.50	26.41	26.32	26.22	26.13
4	21.20	18.00	16.69	15.98	15.52	15.21	14.98	14.80	14.68	14.55	14.37	14.20	14.02	13.93	13.84	13.75	13.65	13.56	13.46
5	16.26	13.27	12.06	11.39	10.97	10.67	10.46	10.29	10.18	10.05	9.89	9.72	9.55	9.47	9.38	9.29	9.20	9.11	9.02
6	13.75	10.92	9.78	9.15	8.75	8.47	8.26	8.10	7.98	7.87	7.72	7.56	7.40	7.31	7.23	7.14	7.06	6.97	6.88
7	12.25	9.55	8.45	7.85	7.46	7.19	6.99	6.84	6.72	6.62	6.47	6.31	6.16	6.07	5.99	5.91	5.82	5.74	5.65
8	11.26	8.65	7.59	7.01	6.63	6.37	6.18	6.03	5.91	5.81	5.67	5.52	5.36	5.28	5.20	5.12	5.03	4.95	4.86
9	10.56	8.02	6.99	6.42	6.06	5.80	5.61	5.47	5.35	5.26	5.11	4.96	4.81	4.73	4.65	4.57	4.48	4.40	4.31
10	10.04	7.56	6.55	5.99	5.64	5.39	5.20	5.06	4.94	4.85	4.71	4.56	4.41	4.33	4.25	4.17	4.08	4.00	3.91
11	9.65	7.21	6.22	5.67	5.32	5.07	4.89	4.74	4.63	4.54	4.40	4.25	4.10	4.02	3.94	3.86	3.78	3.69	3.60
12	9.33	6.93	5.95	5.41	5.06	4.82	4.64	4.50	4.39	4.30	4.16	4.01	3.86	3.78	3.70	3.62	3.54	3.45	3.36
13	9.07	6.70	5.74	5.21	4.86	4.62	4.44	4.30	4.19	4.10	3.96	3.82	3.66	3.59	3.51	3.43	3.34	3.25	3.17
14	8.86	6.51	5.56	5.04	4.69	4.46	4.28	4.14	4.03	3.94	3.80	3.66	3.51	3.43	3.35	3.27	3.18	3.09	3.00
15	8.68	6.36	5.42	4.89	4.56	4.32	4.14	4.00	3.89	3.80	3.67	3.52	3.37	3.29	3.21	3.13	3.05	2.96	2.87
16	8.53	6.23	5.29	4.77	4.44	4.20	4.03	3.89	3.78	3.69	3.55	3.41	3.26	3.18	3.10	3.02	2.93	2.84	2.75
17	8.40	6.11	5.18	4.67	4.34	4.10	3.93	3.79	3.68	3.59	3.46	3.31	3.16	3.08	3.00	2.92	2.83	2.75	2.65
18	8.29	6.01	5.09	4.58	4.25	4.01	3.84	3.71	3.60	3.51	3.37	3.23	3.08	3.00	2.92	2.84	2.75	2.66	2.57
19	8.18	5.93	5.01	4.50	4.17	3.94	3.77	3.63	3.52	3.43	3.30	3.15	3.00	2.92	2.84	2.76	2.67	2.58	2.49
20	8.10	5.85	4.94	4.43	4.10	3.87	3.70	3.56	3.46	3.37	3.23	3.09	2.94	2.86	2.78	2.69	2.61	2.52	2.42
21	8.02	5.78	4.87	4.37	4.04	3.81	3.64	3.51	3.40	3.31	3.17	3.03	2.88	2.80	2.72	2.64	2.55	2.46	2.36
22	7.95	5.72	4.82	4.31	3.99	3.76	3.59	3.45	3.35	3.26	3.12	2.98	2.83	2.75	2.67	2.58	2.50	2.40	2.31
23	7.88	5.66	4.76	4.26	3.94	3.71	3.54	3.41	3.30	3.21	3.07	2.93	2.78	2.70	2.62	2.54	2.45	2.35	2.26
24	7.82	5.61	4.72	4.22	3.90	3.67	3.50	3.36	3.26	3.17	3.03	2.89	2.74	2.66	2.58	2.49	2.40	2.31	2.21
25	7.77	5.57	4.68	4.18	3.85	3.63	3.46	3.32	3.22	3.13	2.99	2.85	2.70	2.62	2.54	2.45	2.36	2.27	2.17
26	7.72	5.53	4.64	4.14	3.82	3.59	3.42	3.29	3.18	3.09	2.96	2.81	2.66	2.58	2.50	2.42	2.33	2.23	2.13
27	7.68	5.49	4.60	4.11	3.78	3.56	3.39	3.26	3.15	3.06	2.93	2.78	2.63	2.55	2.47	2.38	2.29	2.20	2.10
28	7.64	5.45	4.57	4.07	3.75	3.53	3.36	3.23	3.12	3.03	2.90	2.75	2.60	2.52	2.44	2.35	2.26	2.17	2.06
29	7.60	5.42	4.54	4.04	3.73	3.50	3.33	3.20	3.09	3.00	2.87	2.73	2.57	2.49	2.41	2.33	2.23	2.14	2.03
30	7.56	5.39	4.51	4.02	3.70	3.47	3.30	3.17	3.07	2.98	2.84	2.70	2.55	2.47	2.39	2.30	2.21	2.11	2.01
40	7.31	5.18	4.31	3.83	3.51	3.29	3.12	2.99	2.89	2.80	2.66	2.52	2.37	2.29	2.20	2.11	2.02	1.92	1.80
60	7.08	4.98	4.13	3.65	3.34	3.12	2.95	2.82	2.72	2.63	2.50	2.35	2.20	2.12	2.03	1.94	1.84	1.73	1.60
120	6.85	4.79	3.95	3.48	3.17	2.96	2.79	2.66	2.56	2.47	2.34	2.19	2.03	1.95	1.86	1.76	1.66	1.53	1.38
∞	6.63	4.61	3.78	3.32	3.02	2.80	2.64	2.51	2.41	2.32	2.18	2.04	1.88	1.79	1.70	1.59	1.47	1.32	1.00

Table B2. Summary of ANOVA calculations.

Wall Thickness Measurements for Extruder A

Sample	1	2	3	4	5	6	7	8	Total	Average	SS
1	0.142	0.145	0.143	0.14	0.137	0.137	0.139	0.137	0.119	0.14	0.156866
2	0.143	0.146	0.141	0.143	0.136	0.138	0.139	0.141	0.126	0.140875	0.158837
3	0.141	0.145	0.142	0.142	0.135	0.136	0.134	0.139	0.113	0.13925	0.155232
4	0.14	0.143	0.142	0.138	0.137	0.133	0.138	0.139	0.109	0.13875	0.15408
5	0.142	0.145	0.143	0.142	0.136	0.136	0.141	0.141	0.125	0.14075	0.158556
6	0.141	0.144	0.142	0.142	0.137	0.136	0.14	0.14	0.121	0.14025	0.15741
7	0.137	0.142	0.141	0.138	0.138	0.14	0.134	0.14	0.109	0.13875	0.154058
8	0.14	0.142	0.14	0.142	0.135	0.14	0.139	0.139	0.116	0.139625	0.155995
9	0.143	0.146	0.145	0.14	0.135	0.137	0.139	0.138	0.122	0.140375	0.157749
10	0.143	0.147	0.145	0.142	0.142	0.137	0.138	0.141	0.134	0.141875	0.161105
Total	0.125	0.158	0.137	0.122	0.081	0.083	0.094	0.081	−0.093		$X_j^2 =$ 1.569888
Average	0.1412	0.1445	0.1424	0.1409	0.1368	0.137	0.1381	0.1395		0.14005	

Total	SS-Total	0.000767
	SS-Columns	0.000513
	SS-Residual	0.000254
	Mean Square Residual	0.000073
	Residual	0.000003
	F	145.3018

(continued)

169

Table B2. (continued).

Wall Thickness Measurements for Extruder B

Sample	1	2	3	4	5	6	7	8	Total	Average	SS
1	0.146	0.144	0.14	0.134	0.138	0.141	0.145	0.145	1.133	0.141625	0.160583
2	0.161	0.16	0.155	0.151	0.153	0.158	0.163	0.163	1.264	0.158	0.199858
3	0.153	0.159	0.152	0.151	0.153	0.159	0.163	0.156	1.246	0.15575	0.19419
4	0.144	0.146	0.138	0.137	0.139	0.143	0.147	0.141	1.135	0.141875	0.161125
5	0.162	0.159	0.153	0.149	0.152	0.156	0.161	0.161	1.253	0.156625	0.196417
6	0.152	0.154	0.145	0.145	0.148	0.153	0.153	0.149	1.199	0.149875	0.179793
7	0.139	0.139	0.132	0.132	0.134	0.139	0.15	0.137	1.093	0.136625	0.149417
8	0.163	0.16	0.162	0.156	0.152	0.149	0.157	0.158	1.25	0.15625	0.195518
9	0.159	0.158	0.149	0.149	0.151	0.157	0.16	0.157	1.24	0.155	0.192346
10	0.149	0.149	0.143	0.141	0.145	0.149	0.151	0.146	1.173	0.146625	0.172075
Total	0.223	0.223	0.164	0.14	0.16	0.199	0.229	0.208	1.546		
Average	0.1528	0.1528	0.1469	0.1445	0.1465	0.1504	0.1534	0.1513		0.149825	

$X_j^2 =$

Total	1.801322
SS-Total	0.005519
SS-Columns	0.000809
SS-Residual	0.00471
Mean Square Residual	0.000115
Residual	0.000065
F	12.37528

Table B3. Student t ranges for different α values.

$\alpha = 0.10$

ν_E \ P	2	3	4	5	6	7	8	9	10
1	8.93	13.44	16.36	18.49	20.15	21.51	22.64	23.62	24.48
2	4.13	5.73	6.77	7.54	8.14	8.63	9.05	9.41	9.72
3	3.33	4.47	5.20	5.74	6.16	6.51	6.81	7.06	7.29
4	3.01	3.98	4.59	5.03	5.39	5.68	5.93	6.14	6.33
5	2.85	3.72	4.26	4.66	4.98	5.24	5.46	6.56	5.82
6	2.75	3.56	4.07	4.44	4.73	4.97	5.17	5.34	5.50
7	2.68	3.45	3.93	4.28	4.55	4.78	4.97	5.14	5.28
8	2.63	3.37	3.83	4.17	4.43	4.65	4.83	4.99	5.13
9	2.59	3.32	3.76	4.08	4.34	4.54	4.72	4.87	5.01
10	2.56	3.27	3.70	4.02	4.26	4.47	4.64	4.78	4.91
11	2.54	3.23	3.66	3.96	4.20	4.40	4.57	4.71	4.84
12	2.52	3.20	3.62	3.92	4.16	4.35	4.51	4.65	4.78
13	2.50	3.18	3.59	3.88	4.12	4.30	4.46	4.60	4.72
14	2.49	3.16	3.56	3.85	4.08	4.27	4.42	4.56	4.68
15	2.48	3.14	3.54	3.83	4.05	4.23	4.39	4.52	4.64
16	2.47	3.12	3.52	3.80	4.03	4.21	4.36	4.49	4.61
17	2.46	3.11	3.50	3.78	4.00	4.18	4.33	4.46	4.58
18	2.45	3.10	3.49	3.77	3.98	4.16	4.31	4.44	4.55
19	2.45	3.09	3.47	3.75	3.97	4.14	4.29	4.42	4.53
20	2.44	3.08	3.46	3.74	3.95	4.12	4.27	4.40	4.51
24	2.42	3.05	3.42	3.69	3.90	4.07	4.21	4.34	4.44
30	2.40	3.02	3.39	3.65	3.85	4.02	4.16	4.28	4.38
40	2.38	2.99	3.35	3.60	3.80	3.96	4.10	4.21	4.32
60	2.36	2.96	3.31	3.56	3.75	3.91	4.04	4.16	4.25
120	2.34	2.93	3.28	3.52	3.71	3.86	3.99	4.10	4.19
∞	2.33	2.90	3.24	3.48	3.66	3.81	3.93	4.04	4.13

(continued)

Table B3. *(continued).*

$\alpha = 0.05$

ν_E \ P	2	3	4	5	6	7	8	9	10
1	17.97	26.98	32.82	37.08	40.14	43.12	45.40	47.36	49.07
2	6.08	8.33	9.80	10.88	11.74	12.44	13.03	13.54	13.99
3	4.50	5.91	6.82	7.50	8.04	8.48	8.85	9.18	9.46
4	3.93	5.04	5.76	6.29	6.71	7.05	7.35	7.60	7.83
5	3.64	4.60	5.22	5.67	6.03	6.33	6.58	6.80	6.99
6	3.46	4.34	4.90	5.30	5.63	5.90	6.12	6.32	6.49
7	3.34	4.16	4.68	5.06	5.36	5.61	5.82	6.00	6.16
8	3.26	4.04	4.53	4.89	5.17	5.40	5.60	5.77	5.92
9	3.20	3.95	4.41	4.76	5.02	5.24	5.43	5.59	5.74
10	3.15	3.88	4.33	4.65	4.91	5.12	5.30	5.46	5.60
11	3.11	3.82	4.26	4.57	4.82	5.03	5.20	5.35	5.49
12	3.08	3.77	4.20	4.51	4.75	4.95	5.12	5.27	5.39
13	3.06	3.73	4.15	4.45	4.69	4.88	5.05	5.19	5.32
14	3.03	3.70	4.11	4.41	4.64	4.83	4.99	5.13	5.25
15	3.01	3.67	4.08	4.37	4.59	4.78	4.94	5.08	5.20
16	3.00	3.65	4.05	4.33	4.56	4.74	4.90	5.03	5.15
17	2.98	3.63	4.02	4.30	4.52	4.70	4.86	4.99	5.11
18	2.97	3.61	4.00	4.28	4.49	4.67	4.82	4.96	5.07
19	2.96	3.59	3.98	4.25	4.47	4.65	4.79	4.92	5.04
20	2.95	3.58	3.96	4.23	4.45	4.62	4.77	4.90	5.01
24	2.92	3.53	3.90	4.17	4.37	4.54	4.68	4.81	4.92
30	2.89	3.49	3.85	4.10	4.30	4.46	4.60	4.72	4.82
40	2.86	3.44	3.79	4.04	4.23	4.39	4.52	4.63	4.73
60	2.83	3.40	3.74	3.98	4.16	4.31	4.44	4.55	4.65
120	2.80	3.36	3.68	3.92	4.10	4.24	4.36	4.47	4.56
∞	2.77	3.31	3.63	3.86	4.03	4.17	4.29	4.39	4.47

Table B3. (continued).

α = 0.01

ν_E \ P	2	3	4	5	6	7	8	9	10
1	90.03	135.0	164.3	185.6	202.2	215.8	227.2	237.0	245.6
2	14.04	19.02	22.29	24.72	26.63	28.20	29.53	30.68	31.69
3	8.26	10.62	12.17	13.33	14.24	15.00	15.64	16.20	16.69
4	6.51	8.12	9.17	9.96	10.58	11.10	11.55	11.93	12.27
5	5.70	6.98	7.80	8.42	8.91	9.32	9.67	9.97	10.24
6	5.24	6.33	7.03	7.56	7.97	8.32	8.61	8.87	9.10
7	4.95	5.92	6.54	7.01	7.37	7.68	7.94	8.17	8.37
8	4.75	5.64	6.20	6.62	6.96	7.24	7.47	7.68	7.86
9	4.60	5.43	5.96	6.35	6.66	6.91	7.13	7.33	7.49
10	4.48	5.27	5.77	6.14	6.43	6.67	6.87	7.05	7.21
11	4.39	5.15	5.62	5.97	6.25	6.48	6.67	6.84	6.99
12	4.32	5.05	5.50	5.84	6.10	6.32	6.51	6.67	6.81
13	4.26	4.96	5.40	5.73	5.98	6.19	6.37	6.53	6.67
14	4.21	4.89	5.32	5.63	5.88	6.08	6.26	6.41	6.54
15	4.17	4.84	5.25	5.56	5.80	5.99	6.16	6.31	6.44
16	4.13	4.79	5.19	5.49	5.72	5.92	6.08	6.22	6.35
17	4.10	4.74	5.14	5.43	5.66	5.85	6.01	6.15	6.27
18	4.07	4.70	5.09	5.38	5.60	5.79	5.94	6.08	6.20
19	4.05	4.67	5.05	5.33	5.55	5.73	5.89	6.02	6.14
20	4.02	4.64	5.02	5.29	5.51	5.69	5.84	5.97	6.09
24	3.96	4.55	4.91	5.17	5.37	5.54	5.69	5.81	5.92
30	3.89	4.45	4.80	5.05	5.24	5.40	5.54	5.65	5.76
40	3.82	4.37	4.70	4.93	5.11	5.26	5.39	5.50	5.60
60	3.76	4.28	4.59	4.82	4.99	5.13	5.25	5.36	5.45
120	3.70	4.20	4.50	4.71	4.87	5.01	5.12	5.21	5.30
∞	3.64	4.12	4.40	4.60	4.76	4.88	4.99	5.08	5.16

Table B4. Various dimensionless groups and their physical significance.

Dimensionless Group	Symbol	Definition	Significance, Ratio of
Reynolds Number	Re	$\dfrac{\varrho v L}{\mu}$	$\dfrac{\text{Inertial force}}{\text{Viscous force}}$
		ϱ = fluid density	
		v = fluid velocity	
		μ = fluid viscosity	
		L = characteristic dimension	
Froude Number	Fr	$\dfrac{v^2}{Lg}$	$\dfrac{\text{Inertial force}}{\text{Gravitational force}}$
Euler Number	Eu	$\dfrac{p}{\varrho v^2}$	$\dfrac{\text{Pressure}}{2 \times \text{Velocity head}}$
		p = pressure	
Mach Number	Ma	$\dfrac{v}{v_c}$	$\dfrac{\text{Fluid velocity}}{\text{Velocity of sound}}$
Weber Number	We	$\dfrac{\varrho L v^2}{\sigma}$	$\dfrac{\text{Inertial force}}{\text{Surface tension force}}$
		σ = surface tension	
Drag Coefficient	C_D	$\dfrac{(\varrho - \varrho')Lg}{\varrho v^2}$	$\dfrac{\text{Gravitational force}}{\text{Inertial force}}$
		ϱ = density of object	
		ϱ' = density of surrounding fluid	
Fanning Friction Factor	f	$\dfrac{D}{L}\dfrac{\Delta P}{2\varrho v^2}$	$\dfrac{\text{Wall shear stress}}{\text{Velocity head}}$
		D = pipe diameter	
		L = pipe length	

Dimensionless Group	Symbol	Definition	Significance, Ratio of
Nusselt Number (heat transfer)	Nu	$\dfrac{hL}{k}$	$\dfrac{\text{Total heat transfer}}{\text{Conductive heat transfer}}$
		h = heat transfer coefficient	
		k = thermal conductivity	
Prandtl Number	Pr	$\dfrac{C_p\mu}{k}$	$\dfrac{\text{Momentum diffusivity}}{\text{Thermal diffusivity}}$
		C_p = heat capacity	
Peclet Number (heat transfer)	Pe	$\dfrac{C_p\varrho vL}{k} = Re\,Pr$	$\dfrac{\text{Bulk heat transport}}{\text{Conductive heat transfer}}$
Grashof Number	Gr	$\dfrac{gb^3\varrho^2\beta\Delta T}{\mu^2}$	$Re \times \dfrac{\text{Buoyancy force}}{\text{Viscous force}}$
		β = coefficient of expansion	
		ΔT = temperature difference	
		b = height of surface	
Stanton Number	St	$\dfrac{h}{\varrho vC_p} = Nu\,Re^{-1}Pr^{-1}$	$\dfrac{\text{Heat transferred}}{\text{Thermal capacity of fluid}}$
J Factor for Heat Transfer	j_H	$\dfrac{h}{\varrho vC_p}\left(\dfrac{C_p\mu}{k}\right)^{2/3}$	Proportional to $Nu\,Re^{-1}Pr^{-1/3}$
Nusselt Number (mass transfer)	Nu	$\dfrac{k_cL}{\mathscr{D}}$	$\dfrac{\text{Total mass transfer}}{\text{Diffusive mass transfer}}$
		k_c = mass transfer coefficient	
		$\mathscr{D}$ = molecular diffusivity	
Schmidt Number	Sc	$\dfrac{\mu}{\varrho\mathscr{D}}$	$\dfrac{\text{Momentum diffusivity}}{\text{Molecular diffusivity}}$
Peclet Number (mass transfer)	Pe	$\dfrac{Lv}{\mathscr{D}} = Re\,Sc$	$\dfrac{\text{Bulk mass transport}}{\text{Diffusive mass transport}}$
J Factor for Mass Transfer	j_D	$\dfrac{k_c}{v}\left(\dfrac{\mu}{\varrho\mathscr{D}}\right)^{2/3}$	Proportional to $Nu\,Re^{-1}Sc^{-1/3}$

Table **B5.** LOTUS template for linear regression example.

	A	B	C	D	E	F	G	H	M	N	O
1			Linear Regression Model								
2											
3	X	Y	PRED Y	2.5	97.5	deviation	5.00	= degrees of freedom	X^2	Y^2	$X*Y$
4							2.57	= student t			
5	50.00	100.00	71.92	17.45	126.38	28.08	7.00	= n	2500.00	10000.00	5000.00
6	75.00	65.00	65.11	12.58	117.64	−0.11	1064.00	= sum of X	5625.00	4225.00	4875.00
7	100.00	45.00	58.30	7.22	109.39	−13.30	309.00	= sum of Y	10000.00	2025.00	4500.00
8	130.00	30.00	50.13	0.06	100.20	−20.13	208846.00	= sum of X^2	16900.00	900.00	3900.00
9	180.00	27.00	36.52	−13.69	86.73	−9.52	18763.00	= sum of Y^2	32400.00	729.00	4860.00
10	239.00	22.00	20.45	−32.79	73.70	1.55	34193.00	= sum of $X*Y$	57121.00	484.00	5258.00
11	290.00	20.00	6.57	−51.46	64.59	13.43	152.00	= X mean	84100.00	400.00	5800.00
12			85.53	25.91	145.15	−85.53	44.14	= Y mean	0.00	0.00	0.00
13			85.53	25.91	145.15	−85.53	46918.00	= sum of X^2 − ((sum of X)^2)/n	0.00	0.00	0.00
14			85.53	25.91	145.15	−85.53	5122.86	= sum of Y^2 − ((sum of Y)^2)/n	0.00	0.00	0.00

Table B5. *(continued)*.

	A	B	C	D	E	F	G	H	M	N	O
15			85.53	25.91	145.15	−85.53	−12775.00	= sum of X*Y − (sum of X)*(sum of Y)/n	0.00	0.00	0.00
16			85.53	25.91	145.15	−85.53			0.00	0.00	0.00
17			85.53	25.91	145.15	−85.53	−0.27	= slope b0	0.00	0.00	0.00
18			85.53	25.91	145.15	−85.53	85.53	= intercept b1	0.00	0.00	0.00
19			85.53	25.91	145.15	−85.53	−0.82	= r* (coefficient of correlation)	0.00	0.00	0.00
20			85.53	25.91	145.15	−85.53			0.00	0.00	0.00
21			85.53	25.91	145.15	−85.53			0.00	0.00	0.00
22			85.53	25.91	145.15	−85.53	18.14	= st. dev. of points about regression	0.00	0.00	0.00
23			85.53	25.91	145.15	−85.53			0.00	0.00	0.00
24			85.53	25.91	145.15	−85.53		Data ranges:	0.00	0.00	0.00
25			85.53	25.91	145.15	−85.53	50.00	= X min.	0.00	0.00	0.00
26			85.53	25.91	145.15	−85.53	290.00	= X max.	0.00	0.00	0.00
27			85.53	25.91	145.15	−85.53	20.00	= Y min.	0.00	0.00	0.00
28			85.53	25.91	145.15	−85.53	100.00	= Y max.	0.00	0.00	0.00
29			85.53	25.91	145.15	−85.53			0.00	0.00	0.00
30			85.53	25.91	145.15	−85.53			0.00	0.00	0.00

177

Table B6. Example of polynomial fit on LOTUS 1-2-3.

	A	B	C	D	E	F	G	H	I	J	K	L	M
1	X	Y	X^4	X^3	X^2	X	Ycalc						
2	0	89	0	0	0	0	89	Regression Output:					
3	31	89	923521	29791	961	31	89	Constant			89.129611		
4	76	90	33362176	438976	5776	76	91	Std Err of Y Est			1.6007758		
5	100	95	1.0E+08	1000000	10000	100	94	R Squared			0.9947201		
6	159	104	6.4E+08	4019679	25281	159	104	No. of Observations			6		
7	202	113	1.7E+09	8242408	40804	202	113	Degrees of Freedom			1		
8													
9								X Coefficient(s)		$-4.83E-09$	$-2.01E-07$	$1.08E-03$	$-5.24E-02$
10								Std Err of Coef.		$6.07E-08$	0.0000249	0.0032782	0.1453221

178

Table B7. LOTUS template for construction of power-law regression.

Power Law Regression Model

	R	S	T	U	V	W	X	Y	AE	AF	AG
	LN(X)	LN(X)	PRED Y	2.5	97.5	deviation			X^2	Y^2	X*Y
5	3.91	4.61	90.35	86.76	93.94	9.65	7.00	= n	15.30	21.21	18.02
6	4.32	4.17	62.10	58.82	65.38	2.90	34.04	= sum of X	18.64	17.43	18.02
7	4.61	3.81	47.60	44.44	50.75	−2.60	25.37	= sum of Y	21.21	14.49	17.53
8	4.87	3.40	37.34	34.23	40.46	−7.34	167.95	= sum of X^2	23.69	11.57	16.56
9	5.19	3.30	27.64	24.46	30.82	−0.64	94.08	= sum of Y^2	26.97	10.86	17.12
10	5.48	3.09	21.27	17.94	24.59	0.73	121.15	= sum of X*Y	29.99	9.55	16.93
11	5.67	3.00	17.78	14.32	21.25	2.22	4.86	= X mean	32.15	8.97	16.99
12	0.00	0.00	0.00	0.00	9.66	0.00	3.62	= Y mean	0.00	0.00	0.00
13	0.00	0.00	0.00	0.00	9.66	0.00	2.40	= sum X^2 − ((sum of X)^2)/n	0.00	0.00	0.00
14	0.00	0.00	0.00	0.00	9.66	0.00	2.14	= sum Y^2 − ((sum of Y)^2)/n	0.00	0.00	0.00
15	0.00	0.00	0.00	0.00	9.66	0.00	−2.22	= sum X*Y − (sum of X)*(sum of Y)/n	0.00	0.00	0.00
16	0.00	0.00	0.00	0.00	9.66	0.00			0.00	0.00	0.00

(continued)

179

Table B7. *(continued).*

	R	S	T	U	V	W	X	Y	AE	AF	AG
17	0.00	0.00	0.00	0.00	9.66	0.00	−0.92	= slope b0	0.00	0.00	0.00
18	0.00	0.00	0.00	0.00	9.66	0.00	8.12	= intercept b1	0.00	0.00	0.00
19	0.00	0.00	0.00	0.00	9.66	0.00	0.98	= r* (coefficient of correlation)	0.00	0.00	0.00
20	0.00	0.00	0.00	0.00	9.66	0.00			0.00	0.00	0.00
21	0.00	0.00	0.00	0.00	9.66	0.00			0.00	0.00	0.00
22	0.00	0.00	0.00	0.00	9.66	0.00	1.13	= St. Dev. of Points about Regression	0.00	0.00	0.00
23	0.00	0.00	0.00	0.00	9.66	0.00			0.00	0.00	0.00
24	0.00	0.00	0.00	0.00	9.66	0.00			0.00	0.00	0.00
25	0.00	0.00	0.00	0.00	9.66	0.00			0.00	0.00	0.00
26	0.00	0.00	0.00	0.00	9.66	0.00			0.00	0.00	0.00
27	0.00	0.00	0.00	0.00	9.66	0.00			0.00	0.00	0.00
28	0.00	0.00	0.00	0.00	9.66	0.00			0.00	0.00	0.00
29	0.00	0.00	0.00	0.00	9.66	0.00			0.00	0.00	0.00
30	0.00	0.00	0.00	0.00	9.66	0.00			0.00	0.00	0.00

Table **B8.** LOTUS template for exponential fit regression.

	AJ	AK	AL	AM	AN	AO	AT
1	Exponential Model Regression						
2							
3	PRED Y	2.5	97.5	deviation			X*Y
4							
5	70.33	66.42	74.25	29.67			230.26
6	60.29	56.51	64.06	4.71			313.08
7	51.67	48.00	55.34	-6.67			380.67
8	42.95	39.35	46.54	-12.95			442.16
9	31.55	27.94	35.16	-4.55			593.25
10	21.93	18.10	25.76	0.07	3566.93	= sum of X*Y	738.76
11	16.01	11.84	20.18	3.99	3.62	= Y mean	868.76
12	0.00	0.00	0.00	0.00			0.00
13	0.00	0.00	0.00	0.00			0.00
14	0.00	0.00	0.00	0.00			0.00
15	0.00	0.00	0.00	0.00	-289.31	= sum X*Y - (sum of X)*(sum of Y)/n	0.00
16	0.00	0.00	0.00	0.00			0.00
17	0.00	0.00	0.00	0.00	-0.01	= slope b0	0.00
18	0.00	0.00	0.00	0.00	4.56	= intercept b1	0.00
19	0.00	0.00	0.00	0.00	0.91	= r* coefficient of correlation	0.00
20	0.00	0.00	0.00	0.00			0.00
21	0.00	0.00	0.00	0.00			0.00
22	0.00	0.00	0.00	0.00	1.30	= St. Dev. of Points about Regression	0.00
23	0.00	0.00	0.00	0.00			0.00
24	0.00	0.00	0.00	0.00			0.00
25	0.00	0.00	0.00	0.00			0.00
26	0.00	0.00	0.00	0.00			0.00
27	0.00	0.00	0.00	0.00			0.00
28	0.00	0.00	0.00	0.00			0.00

Table B9. Steps to constructing a macro.

(A)	AV3: '/REA5..B200 ~ /REG4 ~		
	Format Label-Prefix Erase Name Justify Protect Unprot. Input		Menu
(B)	AV3: '/REA5..B200 ~ /REG4 ~		Menu
	Create Delete Labels Reset		
	Create or modify a range name		
	AV3: '/REA5..B200 ~ /REG4 ~		Menu
	Enter name:		
(C)	AV3: '/REA5..B200 ~ /REG4 ~		Point
	Enter name: \ E Enter range: AV3		

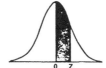

Table B10. Cumulative normal frequency distributions
(area under standard normal curve from 0 to Z).

Z	0.00	0.01	0.02	0.03	0.04	0.05	0.06	0.07	0.08	0.09
0.0	0.0000	0.0040	0.0080	0.0120	0.0160	0.0199	0.0239	0.0279	0.0319	0.0359
0.1	.0398	.0438	.0478	.0517	.0557	.0596	.0636	.0675	.0714	.0753
0.2	.0793	.0832	.0871	.0910	.0948	.0987	.1026	.1064	.1103	.1141
0.3	.1179	.1217	.1255	.1293	.1331	.1368	.1406	.1443	.1480	.1517
0.4	.1554	.1591	.1628	.1664	.1700	.1736	.1772	.1808	.1844	.1879
0.5	.1915	.1950	.1985	.2019	.2054	.2088	.2123	.2157	.2190	.2224
0.6	.2257	.2291	.2324	.2357	.2389	.2422	.2454	.2486	.2517	.2549
0.7	.2580	.2611	.2642	.2673	.2704	.2734	.2764	.2794	.2823	.2852
0.8	.2881	.2910	.2939	.2967	.2995	.3023	.3051	.3078	.3106	.3133
0.9	.3159	.3186	.3212	.3238	.3264	.3289	.3315	.3340	.3365	.3389
1.0	.3413	.3438	.3461	.3485	.3508	.3531	.3554	.3577	.3599	.3621
1.1	.3643	.3665	.3686	.3708	.3729	.3749	.3770	.3790	.3810	.3830
1.2	.3849	.3869	.3888	.3907	.3925	.3944	.3962	.3980	.3997	.4015
1.3	.4032	.4049	.4066	.4082	.4099	.4115	.4131	.4147	.4162	.4177
1.4	.4192	.4207	.4222	.4236	.4251	.4265	.4279	.4292	.4306	.4319
1.5	.4332	.4345	.4357	.4370	.4382	.4394	.4406	.4418	.4429	.4441
1.6	.4452	.4463	.4474	.4484	.4495	.4505	.4515	.4525	.4535	.4545
1.7	.4554	.4564	.4573	.4582	.4591	.4599	.4608	.4616	.4625	.4633
1.8	.4641	.4649	.4656	.4664	.4671	.4678	.4686	.4693	.4699	.4706
1.9	.4713	.4719	.4726	.4732	.4738	.4744	.4750	.4756	.4761	.4767
2.0	.4772	.4778	.4783	.4788	.4793	.4798	.4803	.4808	.4812	.4817
2.1	.4821	.4826	.4830	.4834	.4838	.4842	.4846	.4850	.4854	.4857
2.2	.4861	.4864	.4868	.4871	.4875	.4878	.4881	.4884	.4887	.4890
2.3	.4893	.4896	.4898	.4901	.4904	.4906	.4909	.4911	.4913	.4916
2.4	.4918	.4920	.4922	.4925	.4927	.4929	.4931	.4932	.4934	.4936
2.5	.4938	.4940	.4941	.4943	.4945	.4946	.4948	.4949	.4951	.4952

Table B10. *(continued).*

Z	0.00	0.01	0.02	0.03	0.04	0.05	0.06	0.07	0.08	0.09
2.6	.4953	.4955	.4956	.4957	.4959	.4960	.4961	.4962	.4963	.4964
2.7	.4965	.4966	.4967	.4968	.4969	.4970	.4971	.4972	.4973	.4974
2.8	.4974	.4975	.4976	.4977	.4977	.4978	.4979	.4979	.4980	.4981
2.9	.4981	.4982	.4982	.4983	.4984	.4984	.4985	.4985	.4986	.4986
3.0	.4987	.4987	.4987	.4988	.4988	.4989	.4989	.4989	.4990	.4990
3.1	.4990	.4991	.4991	.4991	.4992	.4992	.4992	.4992	.4993	.4993
3.2	.4993	.4993	.4994	.4994	.4994	.4994	.4994	.4995	.4995	.4995
3.3	.4995	.4995	.4995	.4996	.4996	.4996	.4996	.4996	.4996	.4997
3.4	.4997	.4997	.4997	.4997	.4997	.4997	.4997	.4997	.4997	.4998
3.6	.4998	.4998	.4999	.4999	.4999	.4999	.4999	.4999	.4999	.4999
3.9	.5000									

Table B11. Two-sided Student *t* statistic.

Degrees of Freedom	100 (1 − α)% Confidence Level		
	90%	95%	99%
1	6.314	12.706	63.657
2	2.920	4.303	9.925
3	2.353	3.182	5.841
4	2.132	2.776	4.604
5	2.015	2.571	4.032
6	1.943	2.447	3.707
7	1.895	2.365	3.499
8	1.860	2.306	3.355
9	1.833	2.262	3.250
10	1.812	2.228	3.169
11	1.796	2.201	3.106
12	1.782	2.179	3.055
13	1.771	2.160	3.012
14	1.761	2.145	2.977
15	1.753	2.131	2.947
16	1.746	2.120	2.921
17	1.740	2.110	2.898
18	1.734	2.101	2.878
19	1.729	2.093	2.861
20	1.725	2.083	2.845
21	1.721	2.080	2.831
22	1.717	2.074	2.819
23	1.714	2.069	2.807
24	1.711	2.064	2.797
25	1.708	2.060	2.787

(continued)

Degrees of Freedom	100 (1 − α)% Confidence Level		
	90%	95%	99%
26	1.706	2.056	2.779
27	1.703	2.052	2.771
28	1.701	2.048	2.763
29	1.699	2.045	2.756
30	1.697	2.042	2.750
40	1.684	2.021	2.704
60	1.671	2.000	2.660
120	1.658	1.980	2.617
∞	1.645	1.960	2.576

Table B12. One-sided Student *t* statistic.

Degrees of Freedom	100 (1 − α)% Confidence Level		
	90%	95%	99%
1	3.078	6.314	31.821
2	1.886	2.920	6.965
3	1.638	2.353	4.541
4	1.533	2.132	3.747
5	1.476	2.015	3.365
6	1.440	1.943	3.143
7	1.415	1.895	2.998
8	1.397	1.860	2.896
9	1.383	1.833	2.821
10	1.372	1.812	2.764
11	1.363	1.796	2.718
12	1.356	1.782	2.681
13	1.350	1.771	2.650
14	1.345	1.761	2.624
15	1.341	1.753	2.602
16	1.337	1.746	2.583
17	1.333	1.740	2.567
18	1.330	1.734	2.552
19	1.328	1.729	2.539
20	1.325	1.725	2.528
21	1.323	1.721	2.518
22	1.321	1.717	2.508
23	1.319	1.714	2.500
24	1.318	1.711	2.492
25	1.316	1.708	2.485

Degrees of Freedom	100 $(1 - \alpha)$% Confidence Level		
	90%	95%	99%
26	1.315	1.706	2.479
27	1.314	1.703	2.473
28	1.313	1.701	2.467
29	1.311	1.699	2.462
30	1.310	1.697	2.457
40	1.303	1.684	2.423
60	1.296	1.671	2.390
120	1.289	1.658	2.358
∞	1.282	1.645	2.326

Table B13. Two-sided σ/s statistic.

Degrees of Freedom	100 $(1 - \alpha)$% Confidence Limits					
	90%		95%		99%	
	Lower	Upper	Lower	Upper	Lower	Upper
1	0.510	16.01	0.446	31.94	0.356	160.1
2	0.578	4.407	0.521	6.287	0.434	14.14
3	0.620	2.920	0.567	3.727	0.483	6.468
4	0.649	2.372	0.599	2.875	0.519	4.394
5	0.672	2.090	0.624	2.453	0.546	3.484
6	0.690	1.916	0.644	2.202	0.569	2.979
7	0.705	1.797	0.661	2.035	0.588	2.660
8	0.718	1.711	0.675	1.916	0.604	2.440
9	0.729	1.645	0.688	1.826	0.618	2.277
10	0.739	1.593	0.699	1.755	0.630	2.154
11	0.748	1.551	0.708	1.698	0.641	2.056
12	0.755	1.515	0.717	1.651	0.651	1.976
13	0.762	1.485	0.725	1.611	0.660	1.910
14	0.769	1.460	0.732	1.577	0.669	1.853
15	0.775	1.437	0.739	1.548	0.676	1.806
16	0.780	1.418	0.745	1.522	0.683	1.764
17	0.785	1.400	0.750	1.499	0.690	1.727
18	0.790	1.384	0.756	1.479	0.696	1.695
19	0.794	1.370	0.760	1.461	0.702	1.666
20	0.798	1.358	0.765	1.444	0.707	1.640
22	0.805	1.335	0.773	1.415	0.717	1.595
24	0.812	1.316	0.781	1.391	0.726	1.558
26	0.818	1.300	0.788	1.370	0.734	1.526
28	0.823	1.286	0.794	1.352	0.741	1.499
30	0.828	1.274	0.799	1.337	0.748	1.475
35	0.838	1.248	0.811	1.304	0.762	1.427
40	0.847	1.228	0.821	1.280	0.774	1.390

(continued)

Degrees of Freedom	100 $(1-\alpha)$% Confidence Limits					
	90%		95%		99%	
	Lower	Upper	Lower	Upper	Lower	Upper
45	0.854	1.212	0.829	1.259	0.784	1.361
50	0.861	1.199	0.837	1.243	0.793	1.337
55	0.866	1.188	0.843	1.229	0.801	1.316
60	0.871	1.179	0.849	1.217	0.808	1.299
65	0.875	1.170	0.854	1.207	0.814	1.285
70	0.879	1.163	0.858	1.198	0.820	1.272
80	0.886	1.151	0.866	1.183	0.829	1.250
90	0.892	1.141	0.873	1.171	0.838	1.233
100	0.897	1.133	0.879	1.161	0.845	1.219
120	0.905	1.120	0.888	1.145	0.856	1.196
140	0.911	1.110	0.895	1.133	0.866	1.179
160	0.916	1.102	0.901	1.123	0.873	1.166
180	0.921	1.096	0.906	1.115	0.880	1.155
200	0.925	1.090	0.911	1.109	0.885	1.146
300	0.937	1.072	0.926	1.087	0.904	1.116
400	0.945	1.062	0.935	1.074	0.916	1.099
500	0.951	1.055	0.942	1.066	0.924	1.088
600	0.955	1.050	0.946	1.060	0.930	1.080
800	0.961	1.043	0.953	1.052	0.939	1.068
1000	0.965	1.038	0.958	1.046	0.945	1.061
2000	0.975	1.027	0.970	1.032	0.961	1.042
5000	0.984	1.017	0.981	1.020	0.975	1.026
10000	0.989	1.012	0.986	1.014	0.982	1.019

Table B14. One-sided σ/s statistic.

Lower Confidence Limit				Upper Confidence Limit			
ν	90%	95%	99%	ν	90%	95%	99%
1	.607	.510	.388	1	7.906	16.01	79.06
2	.659	.578	.466	2	3.071	4.407	10.00
3	.693	.620	.514	3	2.264	2.920	5.110
4	.718	.649	.549	4	1.939	2.372	3.671
5	.735	.672	.576	5	1.762	2.090	3.004
6	.752	.690	.597	6	1.651	1.916	2.623
7	.762	.705	.616	7	1.571	1.797	2.377
8	.774	.718	.631	8	1.514	1.711	2.204
9	.783	.729	.645	9	1.470	1.645	2.076
10	.790	.739	.656	10	1.433	1.593	1.977
11	.798	.748	.667	11	1.404	1.551	1.898
12	.803	.755	.677	12	1.380	1.515	1.833
13	.811	.762	.685	13	1.358	1.485	1.779
14	.816	.769	.693	14	1.341	1.460	1.733

Table B14. *(continued).*

	Lower Confidence Limit				Upper Confidence Limit		
ν	90%	95%	99%	ν	90%	95%	99%
15	.819	.775	.700	15	1.324	1.437	1.694
16	.825	.780	.707	16	1.311	1.418	1.659
17	.828	.785	.713	17	1.299	1.400	1.629
18	.833	.790	.719	18	1.287	1.384	1.602
19	.836	.794	.725	19	1.277	1.370	1.578
20	.839	.798	.730	20	1.268	1.358	1.556
22	.845	.805	.739	22	1.252	1.335	1.518
24	.851	.812	.747	24	1.238	1.316	1.487
26	.854	.818	.755	26	1.226	1.300	1.460
28	.860	.823	.762	28	1.216	1.286	1.437
30	.864	.828	.768	30	1.206	1.274	1.416
35	.870	.838	.781	35	1.188	1.248	1.375
40	.877	.847	.792	40	1.174	1.228	1.343
45	.884	.854	.802	45	1.162	1.212	1.318
50	.891	.861	.810	50	1.152	1.199	1.297
55	.894	.866	.818	55	1.143	1.188	1.280
60	.898	.871	.824	60	1.137	1.179	1.265
65	–	.875	.830	65	–	1.170	1.252
70	.905	.879	.835	70	1.125	1.163	1.241
80	.909	.886	.844	80	1.116	1.151	1.222
90	.913	.892	.852	90	1.108	1.141	1.207
100	.921	.897	.858	100	1.102	1.133	1.195
120	.925	.905	.869	120	1.092	1.120	1.175
140	.928	.911	.877	140	1.085	1.110	1.160
160	.933	.916	.884	160	1.078	1.102	1.148
180	.937	.921	.890	180	1.073	1.096	1.139
200	.941	.925	.895	200	1.070	1.090	1.131
300	.949	.937	.913	300	1.056	1.072	1.104
400	.958	.945	.924	400	1.048	1.062	1.089
500	.962	.951	.931	500	1.042	1.055	1.079
600	–	.955	.937	600	–	1.050	1.072
800	–	.961	.945	800	–	1.043	1.062
1000	.971	.965	.950	1000	1.030	1.038	1.055
2000	–	.975	.964	2000	–	1.027	1.038
5000	.990	.984	.977	5000	1.013	1.017	1.024
10000	–	.989	.984	10000	–	1.012	1.017

Table B15. Number of observations for *t*-test of mean.

	Level of *t*-test																			
Single-sided test:	α = 0.005					α = 0.01					α = 0.025					α = 0.05				
Double-sided test:	α = 0.01					α = 0.02					α = 0.05					α = 0.1				
β =	0.01	0.05	0.1	0.2	0.5	0.01	0.05	0.1	0.2	0.5	0.01	0.05	0.1	0.2	0.5	0.01	0.05	0.1	0.2	0.5
Value of δ = (μ − μ₀)/σ																				
0.05																				
0.10																				
0.15																				122
0.20										139					99					70
0.25					110					90				128	64			139	101	45
0.30				134	73				115	63			119	90	45		122	97	71	32
0.35			125	99	58			109	85	47		109	88	67	34		90	72	52	24
0.40		115	97	77	45		101	85	66	37	117	84	68	51	26	101	70	55	40	19
0.45		92	77	62	37	110	81	68	53	30	93	67	54	41	21	80	55	44	33	15
0.50	100	75	63	51	30	90	66	53	43	25	76	54	44	34	18	65	45	36	27	13
0.55	83	63	53	42	26	75	55	46	36	21	63	45	37	28	15	54	38	30	22	11
0.60	71	53	45	36	22	63	47	39	31	18	53	38	32	24	13	46	32	26	19	9
0.65	61	46	39	31	20	55	41	34	27	16	46	33	27	21	12	39	28	22	17	8
0.70	53	40	34	28	17	47	35	30	24	14	40	29	24	19	10	34	24	19	15	8
0.75	47	36	30	25	16	42	31	27	21	13	35	26	21	16	9	30	21	17	13	7
0.80	41	32	27	22	14	37	28	24	19	12	31	22	19	15	9	27	19	15	12	6
0.85	37	29	24	20	13	33	25	21	17	11	28	21	17	13	8	24	17	14	11	6
0.90	34	26	22	18	12	29	23	19	16	10	25	19	16	12	7	21	15	13	10	5
0.95	31	24	20	17	11	27	21	18	14	9	23	17	14	11	7	19	14	11	9	5
1.00	28	22	19	16	10	25	19	16	13	9	21	16	13	10	6	18	13	11	8	5

188

Table B15. *(continued)*.

Level of t-test

Value of $\delta = \dfrac{\mu - \mu_0}{\sigma}$ (Single-sided test / Double-sided test; $\beta =$)	$\alpha = 0.005$ / $\alpha = 0.01$					$\alpha = 0.01$ / $\alpha = 0.02$					$\alpha = 0.025$ / $\alpha = 0.05$					$\alpha = 0.05$ / $\alpha = 0.1$				
	0.01	0.05	0.1	0.2	0.5	0.01	0.05	0.1	0.2	0.5	0.01	0.05	0.1	0.2	0.5	0.01	0.05	0.1	0.2	0.5
1.1	24	19	16	14	9	21	16	14	12	8	18	13	11	9	6	15	11	9	7	
1.2	21	16	14	12	8	18	14	12	10	7	15	12	10	8	5	13	10	8	6	
1.3	18	15	13	11	8	16	13	11	9	6	14	10	9	7		11	8	7	6	
1.4	16	13	12	10	7	14	11	10	9	6	12	9	8	7		10	8	7	5	
1.5	15	12	11	9	7	13	10	9	8	6	11	8	7	6		9	7	6		
1.6	13	11	10	8	6	12	10	9	7	5	10	8	7	6		8	6	6		
1.7	12	10	9	8	6	11	9	8	7		9	7	6	5		8	6	5		
1.8	12	10	9	8	6	10	8	7	7		8	7	6			7	6			
1.9	11	9	8	7	6	10	8	7	6		8	6	6			7	5			
2.0	10	8	8	7	5	9	7	7	6		7	6	5			6				
2.1	10	8	7	7		8	7	6	6		7	6				6				
2.2	9	8	7	6		8	7	6	5		7	6				6				
2.3	9	7	7	6		8	6	6			6	5				5				
2.4	8	7	7	6		7	6	6			6									
2.5	8	7	6	6		7	6	6			6									
3.0	7	6	6	5		6	5	5			5									
3.5	6	5	5			5														
4.0	6																			

Table B16. Number of observations for t-test of difference between two means.

Level of t-test

Single-sided test: → Double-sided test: → Value of $\delta = \dfrac{\mu_1 - \mu_2}{\sigma}$	α = 0.005 α = 0.01 β = 0.01	 0.05	 0.1	 0.2	 0.5	α = 0.01 α = 0.02 0.01	 0.05	 0.1	 0.2	 0.5	α = 0.025 α = 0.05 0.01	 0.05	 0.1	 0.2	 0.5	α = 0.05 α = 0.1 0.01	 0.05	 0.1	 0.2	 0.5
0.05																				
0.10																				
0.15																				
0.20																				137
0.25															124					88
0.30										123					87					61
0.35					110					90					64				102	42
0.40					85					70				100	50			108	78	35
0.45				118	68				101	55			105	79	39		108	86	62	28
0.50				96	55			106	82	45		106	86	64	32		88	70	51	23
0.55			101	79	46		106	88	68	38		87	71	53	27	112	73	58	42	19
0.60		101	85	67	39		90	74	58	32	104	74	60	45	23	89	61	49	36	16
0.65		87	73	57	34	104	77	64	49	27	88	63	51	39	20	76	52	42	30	14
0.70	100	75	63	50	29	90	66	55	43	24	76	55	44	34	17	66	45	36	26	12
0.75	88	66	55	44	26	79	58	48	38	21	67	48	39	29	15	57	40	32	23	11
0.80	77	58	49	39	23	70	51	43	33	19	59	42	34	26	14	50	35	28	21	10
0.85	69	51	43	35	21	62	46	38	30	17	52	37	31	23	12	45	31	25	18	9
0.90	62	46	39	31	19	55	41	34	27	15	47	34	27	21	11	40	28	22	16	8
0.95	56	42	35	28	17	50	37	31	24	14	42	30	25	19	10	36	25	20	15	7
1.00	50	38	32	26	15	45	33	28	22	13	38	27	23	17	9	33	23	18	14	7

190

Table B16. *(continued).*

Level of *t*-test

δ = $\dfrac{\mu_1 - \mu_2}{\sigma}$	Single-sided test: α = 0.005 Double-sided test: α = 0.01					α = 0.01 α = 0.02					α = 0.025 α = 0.05					α = 0.05 α = 0.1				
β =	0.01	0.05	0.1	0.2	0.5	0.01	0.05	0.1	0.2	0.5	0.01	0.05	0.1	0.2	0.5	0.01	0.05	0.1	0.2	0.5
1.1	42	32	27	22	13	38	28	23	19	11	32	23	19	14	8	27	19	15	12	6
1.2	36	27	23	18	11	32	24	20	16	9	27	20	16	12	7	23	16	13	10	5
1.3	31	23	20	16	10	28	21	17	14	8	23	17	14	11	6	20	14	11	9	5
1.4	27	20	17	14	9	24	18	15	12	8	20	15	12	10	6	17	12	10	8	4
1.5	24	18	15	13	8	21	16	14	11	7	18	13	11	9	5	15	11	9	7	4
1.6	21	16	14	11	7	19	14	12	10	6	16	12	10	8	5	14	10	8	6	4
1.7	19	15	13	10	7	17	13	11	9	6	14	11	9	7	4	12	9	7	6	3
1.8	17	13	11	10	6	15	12	10	8	5	13	10	8	6	4	11	8	7	5	
1.9	16	12	11	9	6	14	11	9	8	5	12	9	7	6	4	10	7	6	5	
2.0	14	11	10	8	6	13	10	9	7	5	11	8	7	6	4	9	7	6	4	
2.1	13	10	9	8	5	12	9	8	7	5	10	8	6	5	3	8	6	5	4	
2.2	12	10	8	7	5	11	9	7	6	4	9	7	6	5		8	6	5	4	
2.3	11	9	8	7	5	10	8	7	6	4	9	7	6	5		7	5	5	4	
2.4	11	9	8	6	5	10	8	7	6	4	8	6	5	4		7	5	4	4	
2.5	10	8	7	6	4	9	7	6	5	4	8	6	5	4		6	5	4	3	
3.0	8	6	6	5	4	7	6	5	4	3	6	5	4	4		5	4	3		
3.5	6	5	5	4	3	6	5	4	4		5	4	4	3		4	3			
4.0	6	5	4	4		5	4	4	3		4	4	3			4				

191

Table B17. Critical values of range test for outliers.

Statistic	Number of obs., n	Critical values						
		$\alpha = .30$	$\alpha = .20$	$\alpha = .10$	$\alpha = .05$	$\alpha = .02$	$\alpha = .01$	$\alpha = .005$
$r_{10} = \dfrac{z_{(2)} - z_{(1)}}{z_{(n)} - z_{(1)}}$	3	.684	.781	.886	.941	.976	.988	.994
	4	.471	.560	.679	.765	.846	.889	.926
	5	.373	.451	.557	.642	.729	.780	.821
	6	.318	.386	.482	.560	.644	.698	.740
	7	.281	.344	.434	.507	.586	.637	.680
$r_{11} = \dfrac{z_{(2)} - z_{(1)}}{z_{(n-1)} - z_{(1)}}$	8	.318	.385	.479	.554	.631	.683	.725
	9	.288	.352	.441	.512	.587	.635	.677
	10	.265	.325	.409	.477	.551	.597	.639
$r_{21} = \dfrac{z_{(3)} - z_{(1)}}{z_{(n-1)} - z_{(1)}}$	11	.391	.442	.517	.576	.638	.679	.713
	12	.370	.419	.490	.546	.605	.642	.673
	13	.351	.399	.467	.521	.578	.615	.649
$r_{22} = \dfrac{z_{(3)} - z_{(1)}}{z_{(n-2)} - z_{(1)}}$	14	.370	.421	.492	.546	.602	.641	.674
	15	.353	.402	.472	.525	.579	.616	.647
	16	.338	.386	.454	.507	.559	.595	.624
	17	.325	.373	.438	.490	.542	.577	.605
	18	.314	.361	.424	.475	.527	.561	.589
	19	.304	.350	.412	.462	.514	.547	.575
	20	.295	.340	.401	.450	.502	.535	.562
	21	.287	.331	.391	.440	.491	.524	.551
	22	.280	.323	.382	.430	.481	.514	.541
	23	.274	.316	.374	.421	.472	.505	.532
	24	.258	.310	.367	.413	.484	.497	.524
	25	.262	.304	.380	.406	.457	.489	.516

Table B18. Coefficients for control charts.

Subgroup Size	Control Limit Factors				Estimation of σ*
	Average Chart	Range Chart			
n	A_2	D_3	D_4		d_2
1**	2.660	0	2.267		1.128
2	1.880	0	3.267		1.128
3	1.023	0	3.575		1.693
4	0.729	0	2.282		2.059
5	0.557	0	2.115		2.326
6	0.483	0	2.004		2.534
7	0.419	0.076	1.924		2.704
8	0.373	0.136	1.864		2.847
9	0.337	0.184	1.816		2.970
10	0.308	0.223	1.777		3.078
11	0.285	0.256	1.744		3.173
12	0.266	0.284	1.716		3.258
13	0.249	0.308	1.692		3.336
14	0.235	0.329	1.671		3.407
15	0.223	0.348	1.652		3.472

Control Limits	Averages	Ranges
Upper	$\overline{X} + A_2\overline{R}$	$D_4\overline{R}$
Lower	$\overline{X} - A_2\overline{R}$	$D_3\overline{R}$

*Estimate of $\sigma = \overline{R}/d_2$.

**Individual control charts: use moving ranges to estimate R.

Table B19. Two-sided F distribution.

$\alpha = 0.10$

$\nu_D \backslash \nu_N$	1	2	3	4	5	6	7	8	9	10	12	15	20	24	30	40	60	120	∞
1	161.4	199.5	215.7	224.6	230.2	234.0	236.8	238.9	240.5	241.9	243.9	245.0	248.0	249.1	250.1	251.1	252.2	253.3	254.3
2	18.51	19.00	19.16	19.25	19.30	19.33	19.35	19.37	19.38	19.40	19.41	19.43	19.45	19.45	19.46	19.47	19.48	19.49	19.50
3	10.13	9.55	9.28	9.12	9.01	8.94	8.89	8.85	8.81	8.79	8.74	8.70	8.66	8.64	8.62	8.59	8.57	8.55	8.53
4	7.71	6.94	6.59	6.39	6.26	6.16	6.09	6.04	6.00	5.96	5.91	5.86	5.80	5.77	5.75	5.72	5.69	5.66	5.63
5	6.61	5.79	5.41	5.19	5.05	4.95	4.88	4.82	4.77	4.74	4.68	4.62	4.56	4.53	4.50	4.46	4.43	4.40	4.36
6	5.99	5.14	4.76	4.53	4.35	4.28	4.21	4.15	4.10	4.06	4.00	3.94	3.87	3.84	3.81	3.77	3.74	3.70	3.67
7	5.59	4.74	4.35	4.12	3.97	3.87	3.79	3.73	3.68	3.64	3.57	3.51	3.44	3.41	3.38	3.31	3.30	3.27	3.23
8	5.32	4.46	4.07	3.84	3.69	3.58	3.50	3.44	3.39	3.35	3.28	3.22	3.15	3.12	3.08	3.04	3.01	2.97	2.93
9	5.12	4.26	3.86	3.63	3.48	3.37	3.29	3.23	3.18	3.14	3.07	3.01	2.94	2.90	2.86	2.83	2.79	2.75	2.71
10	4.96	4.10	3.71	3.48	3.35	3.22	3.14	3.07	3.02	2.98	2.91	2.85	2.77	2.74	2.70	2.66	2.62	2.58	2.54
11	4.84	3.98	3.59	3.36	3.20	3.09	3.01	2.95	2.90	2.85	2.79	2.72	2.65	2.61	2.57	2.53	2.49	2.45	2.40
12	4.75	3.89	3.49	3.26	3.11	3.00	2.91	2.85	2.80	2.75	2.69	2.62	2.54	2.51	2.47	2.43	2.38	2.34	2.30
13	4.67	3.81	3.41	3.18	3.03	2.92	2.83	2.77	2.71	2.67	2.60	2.53	2.46	2.42	2.38	2.34	2.30	2.25	2.21
14	4.60	3.74	3.34	3.11	2.96	2.85	2.76	2.70	2.65	2.60	2.53	2.46	2.39	2.35	2.31	2.27	2.22	2.18	2.13
15	4.54	3.68	3.29	3.06	2.90	2.79	2.71	2.64	2.59	2.54	2.48	2.40	2.33	2.29	2.25	2.20	2.16	2.11	2.07
16	4.49	3.63	3.24	3.01	2.85	2.74	2.66	2.59	2.54	2.49	2.42	2.35	2.28	2.24	2.19	2.15	2.11	2.06	2.01
17	4.45	3.59	3.20	2.96	2.81	2.70	2.61	2.55	2.49	2.45	2.38	2.31	2.23	2.19	2.15	2.10	2.06	2.01	1.96
18	4.41	3.55	3.16	2.93	2.77	2.66	2.58	2.51	2.46	2.41	2.34	2.27	2.19	2.15	2.11	2.06	2.02	1.97	1.92
19	4.38	3.52	3.13	2.90	2.74	2.63	2.54	2.48	2.42	2.38	2.31	2.23	2.16	2.11	2.07	2.03	1.98	1.93	1.88
20	4.35	3.49	3.10	2.87	2.71	2.60	2.51	2.45	2.39	2.35	2.28	2.20	2.12	2.08	2.04	1.99	1.95	1.90	1.84
21	4.32	3.47	3.07	2.84	2.68	2.57	2.49	2.42	2.37	2.32	2.25	2.18	2.10	2.05	2.01	1.96	1.92	1.87	1.81
22	4.30	3.44	3.05	2.82	2.66	2.55	2.46	2.40	2.34	2.30	2.23	2.15	2.07	2.03	1.98	1.94	1.89	1.84	1.78
23	4.28	3.42	3.03	2.80	2.61	2.53	2.44	2.37	2.32	2.27	2.20	2.13	2.05	2.01	1.96	1.91	1.86	1.81	1.76
24	4.26	3.40	3.01	2.78	2.62	2.51	2.42	2.36	2.30	2.25	2.18	2.11	2.03	1.98	1.94	1.89	1.84	1.79	1.73
25	4.24	3.39	2.99	2.76	2.60	2.49	2.40	2.34	2.28	2.24	2.16	2.09	2.01	1.96	1.92	1.87	1.82	1.77	1.71
26	4.23	3.37	2.98	2.74	2.59	2.47	2.39	2.32	2.27	2.22	2.15	2.07	1.99	1.95	1.90	1.85	1.80	1.75	1.69
27	4.21	3.35	2.96	2.73	2.57	2.46	2.37	2.31	2.25	2.20	2.13	2.06	1.97	1.93	1.88	1.81	1.79	1.73	1.67
28	4.20	3.34	2.95	2.71	2.56	2.45	2.36	2.29	2.24	2.19	2.12	2.01	1.96	1.91	1.87	1.82	1.77	1.71	1.65
29	4.18	3.33	2.93	2.70	2.55	2.43	2.35	2.28	2.22	2.18	2.10	2.03	1.94	1.90	1.85	1.81	1.75	1.70	1.64
30	4.17	3.32	2.92	2.69	2.53	2.42	2.33	2.27	2.21	2.16	2.09	2.01	1.93	1.89	1.84	1.79	1.74	1.68	1.62
40	4.08	3.23	2.84	2.61	2.45	2.34	2.25	2.18	2.12	2.08	2.00	1.92	1.81	1.79	1.74	1.69	1.64	1.58	1.51
60	4.00	3.15	2.76	2.53	2.37	2.25	2.17	2.10	2.04	1.99	1.92	1.84	1.75	1.70	1.65	1.59	1.53	1.47	1.39
120	3.92	3.07	2.68	2.45	2.29	2.17	2.09	2.02	1.96	1.91	1.83	1.75	1.66	1.61	1.55	1.50	1.43	1.35	1.25
∞	3.84	3.00	2.60	2.37	2.21	2.10	2.01	1.91	1.88	1.83	1.75	1.67	1.57	1.52	1.46	1.39	1.32	1.22	1.00

194

Table B19. *(continued).*

$\alpha = 0.05$

ν_D \ ν_N	1	2	3	4	5	6	7	8	9	10	12	15	20	24	30	40	60	120	∞
1	647.8	799.5	864.2	899.6	921.8	937.1	948.2	956.7	963.3	968.6	976.7	984.9	993.1	997.2	1001	1006	1010	1014	1018
2	38.51	39.00	39.17	39.25	39.30	39.33	39.36	39.37	39.39	39.40	39.41	39.43	39.45	39.46	39.46	39.47	39.48	39.49	39.50
3	17.44	16.04	15.44	15.10	14.88	14.73	14.62	14.54	14.47	14.42	14.34	14.25	14.17	14.12	14.08	14.04	13.99	13.95	13.90
4	12.22	10.65	9.98	9.60	9.36	9.20	9.07	8.98	8.90	8.84	8.75	8.66	8.56	8.51	8.46	8.41	8.36	8.31	8.26
5	10.01	8.43	7.76	7.39	7.15	6.98	6.85	6.76	6.68	6.62	6.52	6.43	6.33	6.28	6.23	6.18	6.12	6.07	6.02
6	8.81	7.26	6.60	6.23	5.99	5.82	5.70	5.60	5.52	5.46	5.37	5.27	5.17	5.12	5.07	5.01	4.96	4.90	4.85
7	8.07	6.54	5.89	5.52	5.29	5.12	4.99	4.90	4.82	4.76	4.67	4.57	4.47	4.42	4.36	4.31	4.25	4.20	4.14
8	7.57	6.06	5.42	5.05	4.82	4.65	4.53	4.43	4.36	4.30	4.20	4.10	4.00	3.95	3.89	3.84	3.78	3.73	3.67
9	7.21	5.71	5.08	4.72	4.48	4.32	4.20	4.10	4.03	3.96	3.87	3.77	3.67	3.61	3.56	3.51	3.45	3.39	3.33
10	6.94	5.46	4.83	4.47	4.24	4.07	3.95	3.85	3.78	3.72	3.62	3.52	3.42	3.37	3.31	3.26	3.20	3.14	3.08
11	6.72	5.26	4.63	4.28	4.04	3.88	3.76	3.66	3.59	3.53	3.43	3.33	3.23	3.17	3.12	3.06	3.00	2.94	2.88
12	6.55	5.10	4.47	4.12	3.89	3.73	3.61	3.51	3.44	3.37	3.28	3.18	3.07	3.02	2.96	2.91	2.85	2.79	2.72
13	6.41	4.97	4.35	4.00	3.77	3.60	3.48	3.39	3.31	3.25	3.15	3.05	2.95	2.89	2.84	2.78	2.72	2.66	2.60
14	6.30	4.86	4.24	3.89	3.66	3.50	3.38	3.29	3.21	3.15	3.05	2.95	2.84	2.79	2.73	2.67	2.61	2.55	2.49
15	6.20	4.77	4.15	3.80	3.58	3.41	3.29	3.20	3.12	3.06	2.96	2.86	2.76	2.70	2.64	2.59	2.52	2.46	2.40
16	6.12	4.69	4.08	3.73	3.50	3.34	3.22	3.12	3.05	2.99	2.89	2.79	2.68	2.63	2.57	2.51	2.45	2.38	2.32
17	6.04	4.62	4.01	3.66	3.44	3.28	3.16	3.06	2.98	2.92	2.82	2.72	2.62	2.56	2.50	2.44	2.38	2.32	2.25
18	5.98	4.56	3.95	3.61	3.38	3.22	3.10	3.01	2.93	2.87	2.77	2.67	2.56	2.50	2.44	2.38	2.32	2.26	2.19
19	5.92	4.51	3.90	3.56	3.33	3.17	3.05	2.96	2.88	2.82	2.72	2.62	2.51	2.45	2.39	2.33	2.27	2.20	2.13
20	5.87	4.46	3.86	3.51	3.29	3.13	3.01	2.91	2.84	2.77	2.68	2.57	2.46	2.41	2.35	2.29	2.22	2.16	2.09
21	5.83	4.42	3.82	3.48	3.25	3.09	2.97	2.87	2.80	2.73	2.64	2.53	2.42	2.37	2.31	2.25	2.18	2.11	2.04
22	5.79	4.38	3.78	3.44	3.22	3.05	2.93	2.84	2.76	2.70	2.60	2.50	2.39	2.33	2.27	2.21	2.14	2.08	2.00
23	5.75	4.35	3.75	3.41	3.18	3.02	2.90	2.81	2.73	2.67	2.57	2.47	2.36	2.30	2.24	2.18	2.11	2.04	1.97
24	5.72	4.32	3.72	3.38	3.15	2.99	2.87	2.78	2.70	2.64	2.54	2.44	2.33	2.27	2.21	2.15	2.08	2.01	1.94
25	5.69	4.29	3.69	3.35	3.13	2.97	2.85	2.75	2.68	2.61	2.51	2.41	2.30	2.24	2.18	2.12	2.05	1.98	1.91
26	5.66	4.27	3.67	3.33	3.10	2.94	2.82	2.73	2.65	2.59	2.49	2.39	2.28	2.22	2.16	2.09	2.03	1.95	1.88
27	5.63	4.24	3.65	3.31	3.08	2.92	2.80	2.71	2.63	2.57	2.47	2.36	2.25	2.19	2.13	2.07	2.00	1.93	1.85
28	5.61	4.22	3.63	3.29	3.06	2.90	2.78	2.69	2.61	2.55	2.45	2.34	2.23	2.17	2.11	2.05	1.98	1.91	1.83
29	5.59	4.20	3.61	3.27	3.04	2.88	2.76	2.67	2.59	2.53	2.43	2.32	2.21	2.15	2.09	2.03	1.96	1.89	1.81
30	5.57	4.18	3.59	3.25	3.03	2.87	2.75	2.65	2.57	2.51	2.41	2.31	2.20	2.14	2.07	2.01	1.94	1.87	1.79
40	5.42	4.05	3.46	3.13	2.90	2.74	2.62	2.53	2.45	2.39	2.29	2.18	2.07	2.01	1.94	1.88	1.80	1.72	1.64
60	5.29	3.93	3.34	3.01	2.79	2.63	2.51	2.41	2.33	2.27	2.17	2.06	1.94	1.88	1.82	1.74	1.67	1.58	1.48
120	5.15	3.80	3.23	2.89	2.67	2.52	2.39	2.30	2.22	2.16	2.05	1.94	1.82	1.76	1.69	1.61	1.53	1.43	1.31
∞	5.02	3.69	3.12	2.79	2.57	2.41	2.29	2.19	2.11	2.05	1.94	1.83	1.71	1.64	1.57	1.48	1.39	1.27	1.00

(continued)

Table B19. (continued).

α = 0.01

ν_D \ ν_N	1	2	3	4	5	6	7	8	9	10	12	15	20	24	30	40	60	120	∞
1	16211	20000	21615	22500	23056	23437	23715	23925	24091	24224	24426	24630	24836	24940	25044	25148	25253	25359	25465
2	198.5	199.0	199.2	199.2	199.3	199.3	199.4	199.4	199.4	199.4	199.4	199.4	199.4	199.5	199.5	199.5	199.5	199.5	199.5
3	55.55	49.80	47.47	46.19	45.39	44.84	44.43	44.13	43.88	43.69	43.39	43.08	42.78	42.62	42.47	42.31	42.15	41.99	41.83
4	31.33	26.28	24.26	23.15	22.46	21.97	21.62	21.35	21.14	20.97	20.70	20.44	20.17	20.03	19.89	19.75	19.61	19.47	19.32
5	22.78	18.31	16.53	15.56	14.94	14.51	14.20	13.96	13.77	13.62	13.38	13.15	12.90	12.78	12.66	12.53	12.40	12.27	12.14
6	18.63	14.54	12.92	12.03	11.46	11.07	10.79	10.57	10.39	10.25	10.03	9.81	9.59	9.47	9.36	9.24	9.12	9.00	8.88
7	16.24	12.40	10.88	10.05	9.52	9.16	8.89	8.68	8.51	8.38	8.18	7.97	7.75	7.65	7.53	7.42	7.31	7.19	7.08
8	14.69	11.04	9.60	8.81	8.30	7.95	7.69	7.50	7.34	7.21	7.01	6.81	6.61	6.50	6.40	6.29	6.18	6.06	5.95
9	13.61	10.11	8.72	7.96	7.47	7.13	6.88	6.69	6.54	6.42	6.23	6.03	5.83	5.73	5.62	5.52	5.41	5.30	5.19
10	12.83	9.43	8.08	7.34	6.87	6.54	6.30	6.12	5.97	5.85	5.66	5.47	5.27	5.17	5.07	4.97	4.86	4.75	4.64
11	12.23	8.91	7.60	6.88	6.42	6.10	5.86	5.68	5.54	5.42	5.24	5.05	4.86	4.76	4.65	4.55	4.44	4.34	4.23
12	11.75	8.51	7.23	6.52	6.07	5.76	5.52	5.35	5.20	5.09	4.91	4.72	4.53	4.43	4.33	4.23	4.12	4.01	3.90
13	11.37	8.19	6.93	6.23	5.79	5.48	5.25	5.08	4.94	4.82	4.64	4.46	4.27	4.17	4.07	3.97	3.87	3.76	3.65
14	11.06	7.92	6.68	6.00	5.56	5.26	5.03	4.86	4.72	4.60	4.43	4.25	4.06	3.96	3.86	3.76	3.66	3.55	3.44
15	10.80	7.70	6.48	5.80	5.37	5.07	4.85	4.67	4.54	4.42	4.25	4.07	3.88	3.79	3.69	3.58	3.48	3.37	3.26
16	10.58	7.51	6.30	5.64	5.21	4.91	4.69	4.52	4.38	4.27	4.10	3.92	3.73	3.64	3.54	3.44	3.33	3.22	3.11
17	10.38	7.35	6.16	5.50	5.07	4.78	4.56	4.39	4.25	4.14	3.97	3.79	3.61	3.51	3.41	3.31	3.21	3.10	2.98
18	10.22	7.21	6.03	5.37	4.96	4.66	4.44	4.28	4.14	4.03	3.86	3.68	3.50	3.40	3.30	3.20	3.10	2.99	2.87
19	10.07	7.09	5.92	5.27	4.85	4.56	4.34	4.18	4.04	3.93	3.76	3.59	3.40	3.31	3.21	3.11	3.00	2.89	2.78
20	9.94	6.99	5.82	5.17	4.76	4.47	4.26	4.09	3.96	3.85	3.68	3.50	3.32	3.22	3.12	3.02	2.92	2.81	2.69
21	9.83	6.89	5.73	5.09	4.68	4.39	4.18	4.01	3.88	3.77	3.60	3.43	3.24	3.15	3.05	2.95	2.84	2.73	2.61
22	9.73	6.81	5.65	5.02	4.61	4.32	4.11	3.94	3.81	3.70	3.54	3.36	3.18	3.08	2.98	2.88	2.77	2.66	2.55
23	9.63	6.73	5.58	4.95	4.54	4.26	4.05	3.88	3.75	3.64	3.47	3.30	3.12	3.02	2.92	2.82	2.71	2.60	2.48
24	9.55	6.66	5.52	4.89	4.49	4.20	3.99	3.83	3.69	3.59	3.42	3.25	3.06	2.97	2.87	2.77	2.66	2.55	2.43
25	9.48	6.60	5.46	4.84	4.43	4.15	3.94	3.78	3.64	3.54	3.37	3.20	3.01	2.92	2.82	2.72	2.61	2.50	2.38
26	9.41	6.54	5.41	4.79	4.38	4.10	3.89	3.73	3.60	3.49	3.33	3.15	2.97	2.87	2.77	2.67	2.56	2.45	2.33
27	9.34	6.49	5.36	4.74	4.34	4.06	3.85	3.69	3.56	3.45	3.28	3.11	2.93	2.83	2.73	2.63	2.52	2.41	2.29
28	9.28	6.44	5.32	4.70	4.30	4.02	3.81	3.65	3.52	3.41	3.25	3.07	2.89	2.79	2.69	2.59	2.48	2.37	2.25
29	9.23	6.40	5.28	4.66	4.26	3.98	3.77	3.61	3.48	3.38	3.21	3.04	2.86	2.76	2.66	2.56	2.45	2.33	2.24
30	9.18	6.35	5.24	4.62	4.23	3.95	3.74	3.58	3.45	3.34	3.18	3.01	2.82	2.73	2.63	2.52	2.42	2.30	2.18
40	8.83	6.07	4.98	4.37	3.99	3.71	3.51	3.35	3.22	3.12	2.95	2.78	2.60	2.50	2.40	2.30	2.18	2.06	1.93
60	8.49	5.79	4.73	4.14	3.76	3.49	3.29	3.13	3.01	2.90	2.74	2.57	2.39	2.29	2.19	2.08	1.96	1.83	1.69
120	8.18	5.54	4.50	3.92	3.55	3.28	3.09	2.93	2.81	2.71	2.54	2.37	2.19	2.09	1.98	1.87	1.75	1.61	1.43
∞	7.88	5.30	4.28	3.72	3.35	3.09	2.90	2.74	2.62	2.52	2.36	2.19	2.00	1.90	1.79	1.67	1.53	1.36	1.00

Table B20. Number of required observations for comparison of two population variances based on F-test[*].

ν	Double-Sided α: 0.02 Single-Sided α: 0.01				Double-Sided α: 0.10 Single-Sided α: 0.05			
	$\beta = 0.01$	$\beta = 0.05$	$\beta = 0.1$	$\beta = 0.5$	$\beta = 0.01$	$\beta = 0.05$	$\beta = 0.1$	$\beta = 0.5$
1	16,420,000	654,200	161,500	4052	654,200	26,070	6,436	161.5
2	9,000	1,881	891.0	99.00	1,881	361.0	171.0	19.00
3	867.7	273.3	158.8	29.46	273.3	86.06	50.01	9.277
4	255.3	102.1	65.62	15.98	102.1	40.81	26.24	6.388
5	120.3	55.39	37.87	10.97	55.39	25.51	17.44	5.050
6	71.67	36.27	25.86	8.466	38.27	18.35	13.09	4.284
7	48.90	26.48	19.47	6.993	26.48	14.34	10.55	3.787
8	36.35	20.73	15.61	6.029	20.73	11.82	8.902	3.428
9	28.63	17.01	13.06	5.351	17.01	10.11	7.757	3.179
10	23.51	14.44	11.26	4.849	14.44	8.870	6.917	2.978
12	17.27	11.16	8.923	4.155	11.16	7.218	5.769	2.687
15	12.41	8.466	6.946	3.522	8.466	5.777	4.740	2.404
20	8.630	6.240	5.270	2.938	6.240	4.512	3.810	2.124
24	7.071	5.275	4.526	2.659	5.275	3.935	3.376	1.984
30	5.693	4.392	3.833	2.386	4.392	3.389	2.957	1.841
40	4.470	3.579	3.183	2.114	3.579	2.866	2.549	1.693
60	3.372	2.817	2.562	1.836	2.817	2.354	2.141	1.534
120	2.350	2.072	1.939	1.533	2.072	1.828	1.710	1.352
	1.000	1.000	1.000	1.000	1.000	1.000	1.000	1.000

[*] The body of this table gives the value of the ratio of the two population variances, σ_1^2/σ_2^2 which remains undetected with probability β if a variance ratio test (i.e., F-test) is used at a significance level of α on the ratio of the estimates of the two variances, s_1^2/s_2^2; each being based on the same number of degrees of freedom (ν).

197

Table B21. Values of Cochran's test for constant variance.

$\alpha = .05$

k \ v	1	2	3	4	5	6	7	8	9	10	16	36	144	∞
2	0.9985	0.9750	0.9392	0.9057	0.8772	0.8534	0.8332	0.8159	0.8010	0.7880	0.7341	0.6602	0.5813	0.5000
3	0.9669	0.8709	0.7977	0.7457	0.7071	0.6771	0.6530	0.6333	0.6167	0.6025	0.5466	0.4748	0.4031	0.3333
4	0.9065	0.7679	0.6841	0.6287	0.5895	0.5598	0.5365	0.5175	0.5017	0.4884	0.4366	0.3720	0.3093	0.2500
5	0.8412	0.6838	0.5981	0.5441	0.5065	0.4783	0.4564	0.4387	0.4241	0.4118	0.3645	0.3066	0.2513	0.2000
6	0.7808	0.6161	0.5321	0.4803	0.4447	0.4184	0.3980	0.3817	0.3682	0.3568	0.3135	0.2612	0.2119	0.1667
7	0.7271	0.5612	0.4800	0.4307	0.3974	0.3726	0.3535	0.3384	0.3259	0.3154	0.2756	0.2278	0.1833	0.1429
8	0.6798	0.5157	0.4377	0.3910	0.3595	0.3362	0.3185	0.3043	0.2926	0.2829	0.2462	0.2022	0.1616	0.1250
9	0.6385	0.4775	0.4027	0.3584	0.3286	0.3067	0.2901	0.2768	0.2659	0.2568	0.2226	0.1820	0.1446	0.1111
10	0.6020	0.4450	0.3733	0.3311	0.3029	0.2823	0.2666	0.2541	0.2439	0.2353	0.2032	0.1655	0.1308	0.1000
12	0.5410	0.3924	0.3264	0.2880	0.2624	0.2439	0.2299	0.2187	0.2098	0.2020	0.1737	0.1403	0.1100	0.0833
15	0.4709	0.3346	0.2758	0.2419	0.2195	0.2034	0.1911	0.1815	0.1736	0.1671	0.1429	0.1144	0.0889	0.0667
20	0.3894	0.2705	0.2205	0.1921	0.1735	0.1602	0.1501	0.1422	0.1357	0.1303	0.1108	0.0879	0.0675	0.0500
24	0.3434	0.2354	0.1907	0.1656	0.1493	0.1374	0.1286	0.1216	0.1160	0.1113	0.0942	0.0743	0.0567	0.0417
30	0.2929	0.1980	0.1593	0.1377	0.1237	0.1137	0.1061	0.1002	0.0958	0.0921	0.0771	0.0604	0.0457	0.0333
40	0.2370	0.1576	0.1259	0.1082	0.0968	0.0887	0.0827	0.0780	0.0745	0.0713	0.0595	0.0462	0.0347	0.0250
60	0.1737	0.1131	0.0895	0.0765	0.0682	0.0623	0.0583	0.0552	0.0520	0.0497	0.0411	0.0316	0.0234	0.0167
120	0.0998	0.0632	0.0495	0.0419	0.0371	0.0337	0.0312	0.0292	0.0279	0.0266	0.0218	0.0165	0.0120	0.0083
∞	0	0	0	0	0	0	0	0	0	0	0	0	0	0

Table B21. (continued).

$\alpha = .01$

ν \ k	1	2	3	4	5	6	7	8	9	10	16	36	144	∞
2	0.9999	0.9950	0.9794	0.9586	0.9373	0.9172	0.8988	0.8823	0.8674	0.8539	0.7949	0.7067	0.6062	0.5000
3	0.9933	0.9423	0.8831	0.8335	0.7933	0.7606	0.7335	0.7107	0.6912	0.6743	0.6059	0.5153	0.4230	0.3333
4	0.9676	0.8843	0.7814	0.7212	0.6761	0.6410	0.6129	0.5897	0.5702	0.5536	0.4884	0.4057	0.3251	0.2500
5	0.9279	0.7885	0.6957	0.6329	0.5875	0.5531	0.5259	0.5037	0.4854	0.4697	0.4094	0.3351	0.2644	0.2000
6	0.8828	0.7218	0.6258	0.5635	0.5195	0.4866	0.4608	0.4401	0.4229	0.4084	0.3529	0.2858	0.2229	0.1667
7	0.8376	0.6644	0.5685	0.5080	0.4659	0.4347	0.4105	0.3911	0.3751	0.3616	0.3105	0.2494	0.1929	0.1429
8	0.7945	0.6152	0.5209	0.4627	0.4226	0.3932	0.3704	0.3522	0.3373	0.3248	0.2779	0.2214	0.1700	0.1250
9	0.7544	0.5727	0.4810	0.4251	0.3870	0.3592	0.3378	0.3207	0.3067	0.2950	0.2514	0.1992	0.1521	0.1111
10	0.7175	0.5358	0.4469	0.3934	0.3572	0.3308	0.3106	0.2945	0.2813	0.2704	0.2297	0.1811	0.1376	0.1000
12	0.6528	0.4751	0.3919	0.3428	0.3099	0.2861	0.2680	0.2535	0.2419	0.2320	0.1961	0.1535	0.1157	0.0833
15	0.5747	0.4069	0.3317	0.2882	0.2593	0.2386	0.2228	0.2014	0.2002	0.1918	0.1612	0.1251	0.0934	0.0667
20	0.4799	0.3297	0.2654	0.2288	0.2048	0.1877	0.1748	0.1646	0.1567	0.1501	0.1248	0.0960	0.0709	0.0500
24	0.4247	0.2871	0.2295	0.1970	0.1759	0.1608	0.1495	0.1406	0.1338	0.1283	0.1060	0.0810	0.0595	0.0417
30	0.3632	0.2412	0.1913	0.1635	0.1454	0.1327	0.1232	0.1157	0.1100	0.1054	0.0867	0.0658	0.0480	0.0333
40	0.2940	0.1915	0.1508	0.1281	0.1135	0.1033	0.0957	0.0898	0.0853	0.0816	0.0668	0.0503	0.0363	0.0250
60	0.2151	0.1371	0.1069	0.0902	0.0796	0.0722	0.0668	0.0625	0.0594	0.0567	0.0461	0.0344	0.0245	0.0167
120	0.1225	0.0759	0.0585	0.0489	0.0429	0.0387	0.0357	0.0334	0.0316	0.0302	0.0242	0.0178	0.0125	0.0083
∞	0	0	0	0	0	0	0	0	0	0	0	0	0	0

Table B22. Percentile values (t_p) for Student t distribution with ν degrees of freedom (shaded area $= p$).

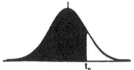

ν	$t_{.995}$	$t_{.99}$	$t_{.975}$	$t_{.95}$	$t_{.90}$	$t_{.80}$	$t_{.75}$	$t_{.70}$	$t_{.60}$	$t_{.55}$
1	63.66	31.82	12.71	6.31	3.08	1.376	1.000	.727	.325	.158
2	9.92	6.96	4.30	2.92	1.89	1.061	.816	.617	.289	.142
3	5.84	4.54	3.18	2.35	1.64	.978	.765	.584	.277	.137
4	4.60	3.75	2.78	2.13	1.53	.941	.741	.569	.271	.134
5	4.03	3.36	2.57	2.02	1.48	.920	.727	.559	.267	.132
6	3.71	3.14	2.45	1.94	1.44	.906	.718	.553	.265	.131
7	3.50	3.00	2.36	1.90	1.42	.896	.711	.549	.263	.130
8	3.36	2.90	2.31	1.86	1.40	.889	.706	.546	.262	.130
9	3.25	2.82	2.26	1.83	1.38	.883	.703	.543	.261	.129
10	3.17	2.76	2.23	1.81	1.37	.879	.700	.542	.260	.129
11	3.11	2.72	2.20	1.80	1.36	.876	.697	.540	.260	.129
12	3.06	2.68	2.18	1.78	1.36	.873	.695	.539	.259	.128
13	3.01	2.65	2.16	1.77	1.35	.870	.694	.538	.259	.128
14	2.98	2.62	2.14	1.76	1.34	.868	.692	.537	.258	.128
15	2.95	2.60	2.13	1.75	1.34	.866	.691	.536	.258	.128
16	2.92	2.58	2.12	1.75	1.34	.865	.690	.535	.258	.128
17	2.90	2.57	2.11	1.74	1.33	.863	.689	.534	.257	.128
18	2.88	2.55	2.10	1.73	1.33	.862	.688	.534	.257	.127
19	2.86	2.54	2.09	1.73	1.33	.861	.688	.533	.257	.127
20	2.84	2.53	2.09	1.72	1.32	.860	.687	.533	.257	.127
21	2.83	2.52	2.08	1.72	1.32	.859	.686	.532	.257	.127
22	2.82	2.51	2.07	1.72	1.32	.858	.686	.532	.256	.127
23	2.81	2.50	2.07	1.71	1.32	.858	.685	.532	.256	.127
24	2.80	2.49	2.06	1.71	1.32	.857	.685	.531	.256	.127
25	2.79	2.48	2.06	1.71	1.32	.856	.684	.531	.256	.127
26	2.78	2.48	2.06	1.71	1.32	.856	.684	.531	.256	.127
27	2.77	2.47	2.05	1.70	1.31	.855	.684	.531	.256	.127
28	2.76	2.47	2.05	1.70	1.31	.855	.683	.530	.256	.127
29	2.76	2.46	2.04	1.70	1.31	.854	.683	.530	.256	.127
30	2.75	2.46	2.04	1.70	1.31	.854	.683	.530	.256	.127
40	2.70	2.42	2.02	1.68	1.30	.851	.681	.529	.255	.126
60	2.66	2.39	2.00	1.67	1.30	.848	.679	.527	.254	.126
120	2.62	2.36	1.98	1.66	1.29	.845	.677	.526	.254	.126
∞	2.58	2.33	1.96	1.645	1.28	.842	.674	.524	.253	.126

Source: R. A. Fisher and F. Yates, *Statistical Tables for Biological, Agricultural and Medical Research* (5th edition), Table III, Oliver and Boyd Ltd., Edinburgh.

Printed and bound by CPI Group (UK) Ltd, Croydon, CR0 4YY

17/10/2024

01775696-0010